T0254883

Data Management in Large-Scale Education Research

Research data management is becoming more complicated. Researchers are collecting more data, using more complex technologies, all the while increasing the visibility of our work with the push for data sharing and open science practices. Ad hoc data management practices may have worked for us in the past, but now others need to understand our processes as well, requiring researchers to be more thoughtful in planning their data management routines.

This book is for anyone involved in a research study involving original data collection. While the book focuses on quantitative data, typically collected from human participants, many of the practices covered can apply to other types of data as well. The book contains foundational context, instructions, and practical examples to help researchers in the field of education begin to understand how to create data management workflows for large-scale, typically federally funded, research studies. The book starts by describing the research life cycle and how data management fits within this larger picture. The remaining chapters are then organized by each phase of the life cycle, with examples of best practices provided for each phase. Finally, considerations on whether the reader should implement, and how to integrate those practices into a workflow, are discussed.

Key Features:

- Provides a holistic approach to the research life cycle, showing how project management and data management processes work in parallel and collaboratively
- Can be read in its entirety, or referenced as needed throughout the life cycle.
- Includes relatable examples specific to education research
- Includes a discussion on how to organize and document data in preparation for data sharing requirements
- Contains links to example documents as well as templates to help readers implement practices

Crystal Lewis is an independent research data management consultant (cghlewis.com). Her experience spans the research life cycle including collecting, curating, sharing, and analyzing data, particularly for federally funded research studies. She is happiest working at the intersection of education research and data management planning, helping researchers build and implement organized processes that lead to more secure, reliable, and usable data.

Data Management in Large-Scale Education Research

Crystal Lewis

CRC Press is an imprint of the
Taylor & Francis Group, an **Informa** business
A CHAPMAN & HALL BOOK

Designed cover image: © Crystal Lewis, Canva Free Content License Agreement

First edition published 2024
by CRC Press
2385 NW Executive Center Drive, Suite 320, Boca Raton FL 33431

and by CRC Press
4 Park Square, Milton Park, Abingdon, Oxon, OX14 4RN

CRC Press is an imprint of Taylor & Francis Group, LLC

© 2025 Crystal Lewis

Reasonable efforts have been made to publish reliable data and information, but the author and publisher cannot assume responsibility for the validity of all materials or the consequences of their use. The authors and publishers have attempted to trace the copyright holders of all material reproduced in this publication and apologize to copyright holders if permission to publish in this form has not been obtained. If any copyright material has not been acknowledged please write and let us know so we may rectify in any future reprint.

Except as permitted under U.S. Copyright Law, no part of this book may be reprinted, reproduced, transmitted, or utilized in any form by any electronic, mechanical, or other means, now known or hereafter invented, including photocopying, microfilming, and recording, or in any information storage or retrieval system, without written permission from the publishers.

For permission to photocopy or use material electronically from this work, access www.copyright.com or contact the Copyright Clearance Center, Inc. (CCC), 222 Rosewood Drive, Danvers, MA 01923, 978-750-8400. For works that are not available on CCC please contact mpkbookspermissions@tandf.co.uk

Trademark notice: Product or corporate names may be trademarks or registered trademarks and are used only for identification and explanation without intent to infringe.

ISBN: 978-1-032-62096-1 (hbk)
ISBN: 978-1-032-62279-8 (pbk)
ISBN: 978-1-032-62283-5 (ebk)

DOI: 10.1201/9781032622835

Typeset in Palatino
by SPi Technologies India Pvt Ltd (Straive)

In memory of my mom, Cassie, who taught me that it's possible to do difficult things.

Contents

About the Author

Crystal Lewis is an independent research data management consultant (cghlewis.com). She has a master's degree in Public Policy from the University of Minnesota, Twin Cities. Her experience spans the research life cycle including collecting, curating, sharing, and analyzing data, particularly for studies funded by federal grants. Crystal is happiest working at the intersection of education research and data management planning, helping researchers build and implement organized processes that lead to more secure, reliable, and usable data. She is a co-organizer for two community groups—R-Ladies St. Louis, an organization focused on promoting gender diversity in the R community, as well as the POWER (Providing Opportunities for Women in Education Research) Data Management Hub, where she facilitates peer data management support in the education research community.

Acknowledgments

This book is a compilation of lessons I have learned in my personal experiences as a data manager, knowledge collected from existing books and papers (many written by librarians or those involved in the open science movement), as well as advice and stories collected through interviews with other researchers who work with data. Unlike research data librarians who are experts in this content, I did not formally study research data management. Instead, much of this book will be based on lessons learned from firsthand experience, and this book is my attempt to hopefully save others from making the same mistakes I have personally made or seen others make. I cannot emphasize enough that if you work for a university and you have the opportunity to consult with a librarian for your project, you absolutely should!

With that said, there is a long list of people I would like to acknowledge for their contributions to this book and for supporting me in this process.

There were many people who graciously allowed me to interview them about their current data management practices. They are Mary McCraken, Ryan Estrellado, Kim Manturuk, Beth Chance, Jessica Logan, Rebecca Schmidt, Sara Hart, and Kerry Shea. These interviews were integral to supplementing my personal knowledge with the broader experience of others in the field. They also affirmed that yes, data management is hard, especially in the context of some of the complicated study designs we work with in education research, and that everyone who works in this field wishes that better training, support systems, and standards existed. Thank-you to everyone who gave me an hour of their time to share their experiences and knowledge! I also have to give a special thank-you to Jessica Logan for being the first person I met who appreciates all things data management as much as I do, and since having our interview, has provided invaluable support while working on this book.

A big thank-you to David Grubbs and Curtis Hill at CRC Press for supporting me through the publishing process. Also, thank-you to everyone who took the time to read and provide feedback on chapters of this book. This includes Meghan Harris, Lexi Swanz, Ally Hanson, Rohan Alexander, Peter Higgins, Emily Riederer, Priyanka Gagneja, Jennifer Huck, Danielle Pico, Kristin Briney, Hope Lancaster, Gizem Solmaz-Ratzlaff, Crystal Steltenpohl, Leigh McLean, Jessica Logan, Chris Schatschneider, Tara Reynolds, Kerry Shea, Joscelin Rocha-Hidalgo, John Muschelli, Sara Hart, and Kyle Husmann. Your revisions and insight helped make this a more cohesive and useful book!

A special thank-you to Keith Herman as well. Many years ago he suggested I write a book titled *Data Management in Large-Scale Education Research,* which seemed unimaginable to me at the time. Thank-you to Keith for envisioning this book long before I ever could.

Also, much appreciation to Wendy Reinke. Joining a project where she had already created documentation and tracking systems was my first glimpse into building tools that help you manage data, and my love of research data management grew out of this experience.

I want to say thank-you to the POWER Issues in Data Management in Education Research Hub. Regularly meeting with this group of data managers, researchers, students, and professors over the last two years has been an amazing source of both support and learning and has greatly increased my understanding of data management.

Last, thank-you to Josh for fully supporting me in the decision to write this book and to Fox for being the reason I remember to step away from my computer from time to time and have fun.

1

Introduction

The results of educational research studies are only as accurate as the data used to produce them.

—Aleata Hubbard (2017)

In 2013, without knowing that the term "research data management" existed, I accepted a position with a prevention science research center. My job was to coordinate the collection and management of data for federally funded randomized controlled trial efficacy studies taking place in K–12 schools, along with a team of investigators, other research staff, part-time data collectors, and graduate students. While I had some experience analyzing and working with education data, i.e., ECLS-K, I had no experience running research grants, collecting original data, or managing research data, but I was excited to learn.

In my time in that position, I learned to plan, schedule, and track data collection activities, create data collection and capture tools, organize and document data inputs, and produce usable data outputs. Yet I didn't learn to do those things through any formal training. There were no books, courses, or workshops that I learned from. I learned from colleagues and a large amount of trial and error. Since then, as I have met more investigators, data managers, and project coordinators in education research, I realize this is a common method for learning data management—mentoring and "winging it". And while learning data management through these informal methods helps us get by, the ramifications of this unstandardized system are felt by both the project team and future data users.

1.1 Why This Book?

Research data management is becoming more complicated. We are collecting more data, in sometimes very novel ways, and using more complex technologies, all the while increasing the visibility of our work with the push for data sharing and open science practices (Briney 2015; Nelson 2022). Ad hoc data management practices may have worked for us in the past, but now

DOI: 10.1201/9781032622835-1

others need to understand our processes as well, requiring researchers to be more thoughtful in planning their data management routines.

1.1.1 Lack of Training, Resources, and Standards

In order to implement thoughtful and standardized data management practices, researchers need training. Yet there is a clear lack of data management training in higher education. In a survey of 274 psychology researchers, Borghi and Van Gulick (2021) found that only 33% of respondents learned data management from college-level coursework, while 64% learned from collaborators, and 52% learned from self-education. In their survey of 202 education researchers (principal investigators and co-principal investigators), Ceviren and Logan (2022) found that over 60% of respondents reported having no formal training in data management, yet across eight different data management practices, respondents were responsible for data management activities anywhere from 25% to 50% of the time. Similarly, in a survey of 150 graduate students in a school of education, when asked if they needed more training in research data management, the average overall score on a scale from 1 to 100 was 80, while the overall confidence in managing data score was only 40 (Zhou, Xu, and Kogut 2023). Furthermore, of the training that does exist, usually provided through university library systems, most material is either discipline agnostic or STEM focused, leaving a gap in training on how to apply skills to the field of education which has unique issues, particularly around working with human subjects data (Thielen and Hess 2017).

Without training, resources and formal support systems are the next best option for learning best practices. Within university systems, in addition to providing periodic training, research data librarians provide data management planning consultation for researchers and their teams. There is also a wealth of existing research data management resources written for broad audiences which I will reference in this book. However, while education researchers are starting to put out some excellent resources (Neild, Robinson, and Agufa 2022; Reynolds, Schatschneider, and Logan 2022), I still find there is a dearth of practical guides for researchers to refer to when building a data management workflow in the field of education, especially those working on large-scale longitudinal research grants where there are many moving pieces. Researchers are often collecting data in real-world environments, such as school systems, and keeping that data secure and reliable in a deliberate and orderly way can be overwhelming.

Last, unfortunately, while other fields of research, such as psychology, appear to be banding together to develop standards around how to structure and document data (Kline et al. 2018), the field of education has yet to develop shared rules for things such as data documentation or data formats. This lack of standards leads to inconsistencies in the quality and usability of data products across the field (Borghi and Van Gulick 2022).

1.1.2 Consequences

A lack of training in data management practices and an absence of agreed-upon standards in the field of education lead to consequences. Implementing subpar and inconsistent data management practices, while typically only resulting in frustration and time lost, also has the potential to be devastating, resulting in analyzing erroneous data or even unusable or lost data. In a review of 1,082 retracted publications from the journal *PubMed* from 2013 to 2016, authors found that 32% of retractions were due to data management errors (Campos-Varela and Ruano-Raviña 2019). In a 2013 study surveying 360 graduate students about their data management practices, 14% of students indicated they had to recollect data that had been previously collected because they could not find a file or the file had been corrupted, while 17% of students said they had lost a file and been unable to recollect it (Doucette and Fyfe 2013). In their study of 488 researchers who had published in a psychology journal between 2010 and 2018, Kovacs et al. (2021) asked respondents about their data management mistakes and found that the most serious data management mistakes reported led to a range of consequences including time loss, frustration, and even erroneous conclusions.

Poor data management can even prevent researchers from implementing other good open science practices. In waves 1 and 2 of the Open Scholarship Survey being collected by the Center for Open Science, the team has found that of the education researchers surveyed who are currently not publicly sharing their research data, approximately 15% mentioned "being nervous about mistakes" as a reason for not sharing (Beaudry et al. 2022). Similarly, when surveying 780 researchers in the field of psychology, researchers found that 38% of respondents agreed that a "fear of discovery of errors in the data" posed a barrier to data sharing (Houtkoop et al. 2018).

The well-known replication crisis is another reason to be concerned with data management. Failure to implement practices such as quality documentation or standardization of practices (among many other reasons), resulted in one study finding that across 1,500 researchers surveyed, more than 70% had tried and failed to reproduce another researcher's study (Baker 2016).

1.2 About This Book

While the field of education may not have agreed-upon guidelines for data management, there are still practices that are proven to result in more secure, reproducible, and reliable data. My hope is that this book can be a foundation to help researchers think through how to build a quality, standardized data management workflow that works for their team and their projects. As suggested in the title of this book, this content is designed to specifically help teams navigate the complicated workflows associated with large-scale research, such as randomized controlled trial studies, but ultimately these practices are applicable to any research project, no matter the scale.

If this is your first time opening this book, I recommend reading this book from cover to cover. Much of the information in later chapters builds off of content from earlier chapters. With that said, once you have an understanding of what is contained in each chapter, this book is absolutely meant to become a handbook to be referenced as needed when you are ready to start planning a specific phase of your project.

1.2.1 What This Book Will Cover

This book begins, like many other books in this subject area, by describing the research life cycle and how data management fits within the larger picture. The remaining chapters are then organized by each phase of the life cycle, with examples of best practices provided for each phase. Considerations on whether you should implement and how to integrate those practices into your workflow will be discussed.

Links to templates, checklists, and example documents are provided throughout this book. If you prefer clickable links, you can view the online, open access version of this book at https://datamgmtinedresearch.com/.

1.2.2 What This Book Will Not Cover

It is important to also point out what this book will not cover. This book is intended to be tool agnostic and provide suggestions that anyone can use, no matter what tools you work with, especially when it comes to data cleaning. Therefore, while I might mention options of tools you can use for different tasks, I will not advocate for any specific tools.

There are also no specific coding practices or syntax included in this book. In many ways I feel that the actual "data cleaning" phase of data management is the *easiest* phase to implement, as long as you implement good practices up until that point. Because of that, this book introduces practices in all phases leading up to data cleaning that will prepare your data for minimal cleaning. With that said, I do provide examples of what I would expect to see in a data cleaning process, I just do not provide steps for any specific software system. That is beyond the scope of this book.

This book will also not talk about analysis or preparing data for analysis through means such as data imputation, removal of legitimate outliers, or calculating analysis-specific variables. Written from the perspective of a data manager, the end goal of data management is to build datasets for general data sharing. This means we will cover practices that keep data in its most complete and true, but usable, form, for any future researcher to analyze in a way that works best for them.

Last, I want to acknowledge that education research studies, while all happening under a similar umbrella of study, are each unique in their design and requirements. It would be impossible for me to provide examples throughout this book that are applicable to every type of project a reader may encounter. Instead, I have done my best to provide examples that I think are generally relatable to a wide audience of researchers, in hopes that you can then extrapolate those examples to your own specific work.

1.3 Who This Book Is For

This book is for anyone involved in a research study involving original data collection. In particular, this book focuses on quantitative data, typically collected from human participants, although many of the practices covered could apply to other types of data as well. This book also applies to any team member, ranging from investigators, to data managers, to project staff, to students, to contractual data collectors. The contents of this book are useful for anyone who may have a part in planning, collecting, or organizing research study data.

1.4 Final Note

Planning and implementing new data management practices on top of planning the implementation of your entire research grant can feel overwhelming. However, the idea of this book is to find the practices that work for you and your team and implement them consistently. For some teams that may look like implementing just a few of the suggestions mentioned; for others it may involve implementing all of them. Improving your data management workflow is a process and it becomes easier over time as those practices become part of your normal routine. At some point you may even find that you enjoy working on data management processes as you start to see the benefits of their implementation!

2

Research Data Management Overview

2.1 What Is Research Data Management?

Research data management (RDM) involves the organization, storage, preservation, and dissemination of research study data (Borghi and Van Gulick 2022). Research study data includes materials generated or collected throughout a research process (National Endowment for the Humanities 2018). As you can imagine, this broad definition includes much more than just the management of digital datasets. It also includes physical files, documentation, artifacts, recordings, and more. RDM is a substantial undertaking that begins long before data are ever collected, during the planning phase, and continues well after a research project ends during the archiving and sharing phase.

2.2 Data Management Standards

It's important for research data management practices to be structured around standards—rules for how data should be collected, formatted, described, and shared (Borghi and Van Gulick 2022; Koos 2023). Implementing standards for procedures such as how variables should be collected and named, which items from common measures should be shared, and how data should be formatted and documented, leads to more findable and usable data within fields and provides the added benefit of allowing researchers to integrate datasets without painstaking work to harmonize the data.

Some fields have adopted standards across the research life cycle, such as CDISC standards used by clinical researchers (CDISC 2023), other fields have adopted standards specifically around metadata, such as the TEI standards used in digital humanities (Burnard 2014) or the ISO 19115 standard used for geospatial data (Michener 2015), and through grassroots efforts, other fields such as psychology are developing their own standards for things such as

DOI: 10.1201/9781032622835-2

data formatting and documentation (Kline et al. 2018) based on the FAIR principles and inspired by the BIDS standard (BIDS-Contributors 2022). Yet, it is common knowledge that there are currently no agreed-upon norms in the field of education research (Institute of Education Sciences 2023a; Logan and Hart 2023). The rules for how to collect, format, and document data are often left up to each individual team, as long as external compliance requirements are met (Tenopir et al. 2016).

Without agreed-upon standards in the field, it is important for research teams to develop their own data management standards that apply within and across all of their projects. Developing internal standards, and applying them in a reproducible data management workflow, allows practices to be implemented consistently and with fidelity.

2.3 Why Care about Research Data Management?

There are both external and personal reasons to care about developing research data management standards for your projects.

2.3.1 External Reasons

1. **Funder compliance**: Researchers applying for federal funding will be required to submit a data management plan (see Chapter 5) along with their grant proposal (Holdren 2013; Nelson 2022). The contents of these plans may vary slightly across agencies, but the shared purpose of these documents is to facilitate good data management practices and to mandate open sharing of data to maximize scientific outputs and benefits to society. Along with this mandatory data sharing policy comes the incentive to manage your data for the purposes of data sharing (Borghi and Van Gulick 2022).
2. **Journal compliance**: Depending on what journal you publish with, providing open access to the data associated with your publication may be a requirement (see *PLOS ONE* (https://journals.plos.org/plosone/) and *AMPPS* (https://www.psychologicalscience.org/publications/ampps) as examples). Again, along with data sharing, comes the incentive to manage your data in a thoughtful, responsible, and organized way.
3. **Compliance with mandates**: Depending on your research design and the sensitivity level of the data you are collecting (see Section 4.2), there are a variety of policies as well as legal or contractual obligations you may need to consider when managing data (see Section 4.3). If you

are required to submit your project to an Institutional Review Board (IRB), the board will review and monitor your data management practices. Concerned with the welfare, rights, and privacy of research participants, your IRB will have rules for how data is securely collected, managed, and shared (Filip 2023). Your data may also be subject to laws, such as HIPAA or FERPA, which regulate the privacy and exchange of personal information. If working with research partners, you may also need to monitor and honor any conditions laid out in data sharing or other legal agreements. Additionally, your organization may have their own institutional data policies that mandate how data must be cared for and secured.

4. **Open science practices**: With a growing interest in open science practices, sharing well-managed and documented data helps to build trust in the research process (Renbarger et al. 2022). Sharing data that is curated in a reproducible way is "a strong indicator to fellow researchers of rigor, trustworthiness, and transparency in scientific research" (Alston and Rick 2021, 2). It also allows others to replicate and learn from your work, validate your results to strengthen evidence, as well as potentially catch errors in your work, preventing decisions being made based on incorrect data. Sharing your data with sufficient documentation and standardized metadata can also lead to more collaboration and greater impact as collaborators are able to access and understand your data with ease (Borghi and Van Gulick 2022; Eaker 2016).
5. **Data management is a matter of ethics**: In education research we are often collecting data from human participants. As a result, data management is an ethical issue. It is our responsibility to have well-designed research studies with data collection, management, ownership, and sharing practices that consider the environmental, social, cultural, historical, and political context of the data we are working with (Alexander 2023). Furthermore, collecting data from human participants means people are giving their time and energy and entrusting us with their information. Implementing poor data management that leads to irrelevant, unusable, or compromised data is a huge disservice to research participants and erodes trust in the research process (Feeney, Kopper, and Sautmann 2022; Gammons 2022).

2.3.2 Personal Reasons

There are also many compelling personal reasons to manage your data in a reproducible and standardized way.

1. **Reduces data curation debt**: Taking the time to plan and implement quality data management throughout the entire research study reduces data curation debt caused by suboptimal data management practices (Butters, Wilson, and Burton 2020). Having poorly collected, managed, or documented data may make your data unusable, either permanently or until errors are corrected. Decreasing or removing this debt reduces the time, energy, and resources spent possibly recollecting data or scrambling at the end of your study to get your data up to acceptable standards.
2. **Facilitates use of your data**: Every member of your research team being able to find and understand your project data and documentation is a huge benefit. It allows for the easy use and reuse of your data and hastens efforts like the publication process (Markowetz 2015). Not having to search around for participant numbers or having to ask which version of a file to use allows your team to spend more time analyzing and less time playing detective.
3. **Encourages validation**: Implementing reproducible data management practices encourages and allows your team to internally replicate and validate your processes to ensure your outputs are accurate (Alston and Rick 2021).
4. **Improves continuity**: Data management practices, such as documentation, ensure implementation fidelity during your project. This includes consistently implementing practices during a longitudinal project or across sites. It also improves project continuity through staff turnover. Having thoroughly documented procedures allows new staff to pick up right where the former staff member left off and implement the project with fidelity (Borghi and Van Gulick 2021; Princeton University 2023b). Furthermore, good data management enables continuity when handing off projects to collaborators or when picking up your own projects after a long hiatus (Markowetz 2015).
5. **Increases efficiency**: Documenting and automating data management tasks reduces duplication of efforts for repeating tasks, especially in longitudinal studies.
6. **Upholds research integrity**: Errors come in many forms, from both humans and technology (Kovacs, Hoekstra, and Aczel 2021; Strand 2021). We've seen evidence of this in the papers cited as being retracted for "unreliable data" in the blog Retraction Watch (https://retractionwatch.com/). Implementing quality assurance and control procedures reduces the chances of errors occurring and allows you to have confidence in your data. Without implementing these

practices, your research findings could include extra noise, missing data, or erroneous or misleading results.

7. **Improves data security**: Quality data management and preservation practices reduce the risk of lost or stolen data, the risk of data becoming corrupted or inaccessible, and the risk of breaking confidentiality agreements.

2.4 Existing Frameworks

Data management does not live in a space all alone. It coexists with other frameworks that impact how and why data are managed, and it is important to be familiar with them as they will provide a foundation for you as you build your data management structures.

2.4.1 FAIR

In 2016, the FAIR Principles were published in *Scientific Data* (Wilkinson et al. 2016), outlining four guiding principles for scientific data management and stewardship. These principles were created to improve and support the reuse of scholarly data, specifically the ability of machines to access and read data. They are the foundation for how all digital data should be publicly shared. The principles are:

F: Findable

All data should be findable through a persistent identifier and have thorough, searchable metadata. These practices aid in the long-term discovery of information and provide registered citations.

A: Accessible

Users should be able to access your data. This can mean your data is available in a repository or through a request system. At minimum, a user should be able to access the metadata, even if the actual data are not openly available.

I: Interoperable

Your data and metadata should use standardized vocabularies as well as formats. Both humans and machines should be able to read and interpret your data. Licenses should not pose a barrier to usage. Data should be available in open formats that can be accessed by any software (e.g., CSV, TXT, DAT).

R: Reusable

To provide context for the reuse of your data, your metadata should give insight into data provenance, providing a project description, an overview of the data workflow, as well as what authors to cite for appropriate attribution. You should also have clear licensing for data use.

2.4.2 SEER

The SEER principles, developed in 2018 by the Institute of Education Sciences (IES), provide Standards for Excellence in Education Research (Institute of Education Sciences 2022). While the principles broadly cover the entire life cycle of a research study, they provide context for good data management within an education research study. The SEER principles include:

- Preregister studies
- Make findings, methods, and data open
- Identify interventions' core components
- Document treatment implementation and contrast
- Analyze interventions' costs
- Focus on meaningful outcomes
- Facilitate generalization of study findings
- Support scaling of promising results

2.4.3 Open Science

The concept of open science has pushed quality data management to the forefront, bringing visibility to its cause, as well as advances in practices and urgency to implement them. Open science aims to make scientific research and dissemination accessible for all, making the need for good data management practices absolutely necessary. Open science advocates for transparent and reproducible practices through means such as open data, open analysis, open materials, preregistration, and open access (Van Dijk, Schatschneider, and Hart 2021). Organizations, such as the Center for Open Science (https://www.cos.io), have become a well-known proponent of open science, offering the Open Science Framework (OSF) (Foster and Deardorff 2017) as a tool to promote open science through the entire research life cycle. Furthermore, many education funders have aligned their requirements with these open science practices, such as openly sharing study data and preregistration of study methods.

NOTE

When working with specific populations, there may be other principles to consider that complement FAIR principles and open science practices and provide further guidance for working with and protecting data collected from those specific communities. As an example, when conducting research with Indigenous populations, it is important to consider Indigenous data sovereignty which recognizes the rights of Indigenous peoples to own, control, access, and use data collected about their communities and lands, and to engage Indigenous communities when planning data management for your study.

(Carroll et al. 2020; National Institutes of Health 2022)

2.5 Terminology

Before moving forward in this book, it is important to have a shared understanding of terminology used. Many concepts in education research have synonymous terms that can be used interchangeably. Across different institutions, researchers may use all or some of these terms. Please review the Glossary to gain a better understanding of how terms will be used throughout this book.

2.6 The Research Life Cycle

The remainder of this book will be organized into chapters that provide best practices for each phase of the research project life cycle. It is imperative to understand this life cycle in order to see the flow of data through a project, as well as to understand how everything in a project is connected. If phases are skipped, the whole project will suffer.

In Figure 2.1, you can see how throughout the project, data management roles and project coordination roles work in parallel and collaboratively. These teams may be made up of the same people or different members, but either way, both workflows must happen, and they must work together.

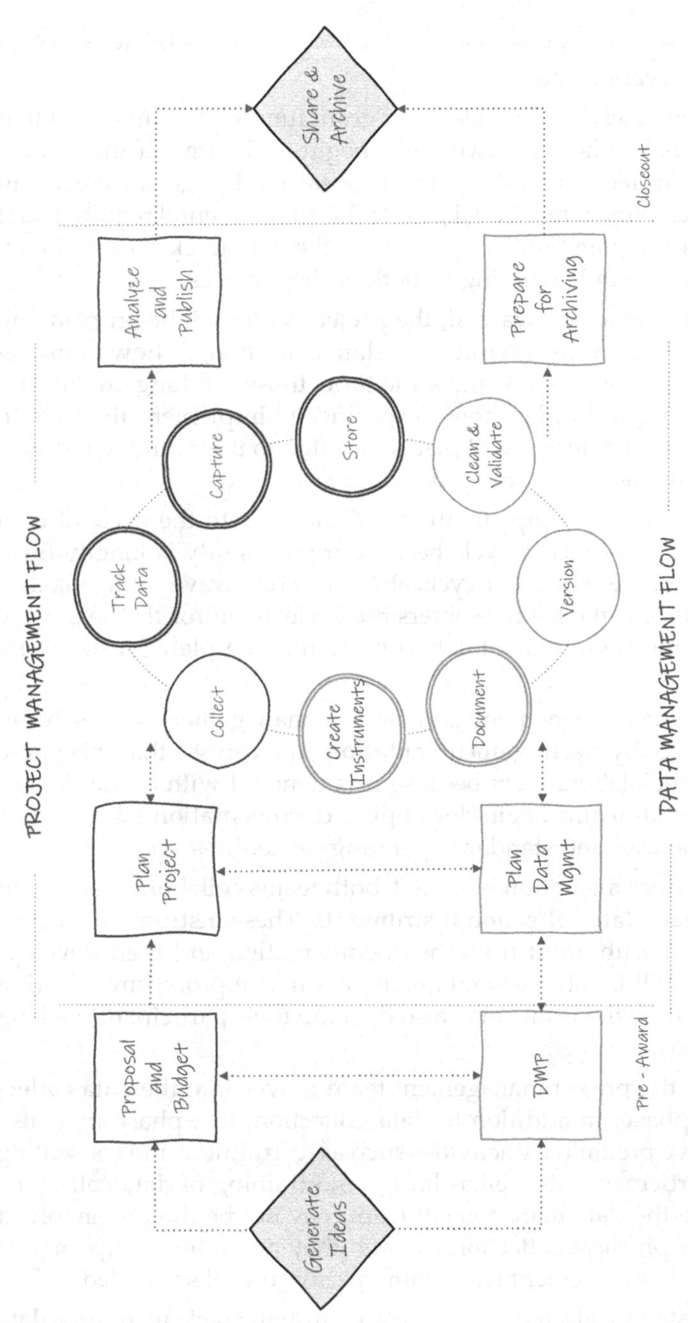

FIGURE 2.1
The research project life cycle.

Let's walk through this chart.

1. In a typical study, a team first begins by **generating ideas**, deciding what they want to study.
2. Then, most likely, they will look for grant funding to implement that study. This is where the two paths begin to diverge. If the team is applying for federal funding, the proposal and budget are created in the project management track, while the supplemental required **data management plan** (DMP) is created in the data track. Again, it may be the same people working on both of these pieces.
3. Next, if the grant is awarded, the project team will begin planning things such as hiring, recruitment, data collection, or how to implement the intervention. At the same time, those working on the data team will begin to **plan** how to specifically implement the two- to five-page data management plan submitted to their funder and start putting any necessary structures into place.
4. Once planning is complete, the team moves into the cycle of data collection. It is called a cycle because if your study is longitudinal, every step here will occur cyclically. Once one wave of data collection wraps up, the team re-enters the cycle again for the next wave of data collection, until all data collection is complete for the entire project.
    - The data management and project management teams begin the cycle by starting **documentation**. You can see that this phase occurs collaboratively because it is denoted with a double outline. Both teams begin developing documentation such as data dictionaries and standard operating procedures.
    - Once documentation is started, both teams collaboratively begin to **create data collection instruments**. These instruments will be created with input from the documentation and their development will involve several quality assurance procedures. During this phase the teams may also develop their participant tracking database.
    - Next, the project management team moves into the **data collection** phase. In addition to data collection, this phase may also involve preliminary activities such as recruitment and consenting of participants, as well as hiring and training of data collectors. While the data management team may not be directly involved in the physical collection of data, they continue to support the project management team with quality control as needed.
    - As data is collected, the project team will **track incoming data** in the participant tracking database. The data management team

will collaborate with the project management team to help troubleshoot anything related to the tracking database or any issues discovered with the data during tracking.

- Next, once data is collected, the teams move into the **data capture** phase. This is where teams are actively retrieving electronically collected data or converting paper data into a digital format. Oftentimes, this again is a collaborative effort between the project management team and the data team.
- Once the data is captured, files need to be **stored**. While the data team may be in charge of setting up and monitoring the storage efforts, the project team may be the ones actively retrieving and storing the data.
- Next the teams move into the **cleaning and validation** phase. At this time the data team is reviewing data cleaning plans, writing data cleaning scripts, and actively cleaning data from the most recent data collection round.
- Last, the data team will create and save new **versions** of data as it is updated, or errors are found.

5. The teams then only move out of the active data collection cycle when all data collection for the project is complete. At this time the project team begins analyzing study data and working on publications as well as any final grant reports. They are able to do this because of the organized processes implemented during the data collection cycle. Since data was managed and cleaned throughout, data is ready for analysis as soon as data collection is complete. Then, while the project team is analyzing data, the data team is doing any additional **preparation to archive** data for long-term storage and public sharing.
6. Last, as the grant is closing out, the team submits data for **public sharing**.

As you work through the remaining chapters of this book, Figure 2.1 will be a guide to navigating where each phase of practices fits into the larger picture. If at any point you feel overwhelmed by the information in this book, the Appendix provides a summary of some of the most common activities that occur in each phase. This digestible list can be a helpful reminder that all of this information can be boiled down to achievable tasks that are implemented over an extended period of time.

3

Data Organization

Before jumping into the project life cycle, we need to have a basic understanding of what data looks like. Understanding the basics of data organization helps us to make informed decisions throughout the life cycle that will result in clear, analyzable information.

3.1 Basics of a Dataset

In education research, data is often collected internally by a team using an instrument such as a questionnaire, an observation form, an interview guide, or an assessment. However, data may also be collected by external entities, such as districts, states, or other agencies.

Those data come in many forms (e.g., video, transcripts, documents, data files), represented as text, numbers, or multimedia (USGS 2023). In the world of quantitative education research, we are often working with digital data in the form of a dataset, a structured collection of data. A dataset is organized in a rectangular format which allows the information to be machine-readable. Rectangular, also called tabular, datasets are made up of columns and rows (see Figure 3.1).

3.1.1 Columns

The columns in your dataset will consist of the following types of variables:

- Variables you collect
  - These are variables collected from an instrument or external source.
- Variables you create
  - These may be indicators you create (e.g., cohort, treatment, time).
  - Or they me be variables derived for summary purposes (e.g., means, sum scores).

DOI: 10.1201/9781032622835-3

stu_id	item1	item2	item3
12412	3	2	1
12419	4	1	5
12428	-99	3	2

Columns

Rows

FIGURE 3.1
Basic format of a dataset.

- Identifier variables
  - You must also include values that uniquely identify subjects in your data (e.g., a student unique identifier).
  - See Section 10.4 for more information on creating unique identifier variables.

3.1.1.1 Column Attributes

The columns, or variables, in your dataset also have the following attributes:

1. Variable names
   - A variable name is the short representation of the information contained in a column.
   - Variable names must be unique. No variable name in a dataset can repeat. We will talk more about variable naming when we discuss style guides in Chapter 9.
2. Variable types
   - A variable's type determines allowable values for a variable, the operations that can be performed on the variable, and how the values are stored.
   - Example types include numeric, character (also called text or string), or date. Types can also be more narrowly defined as needed (e.g., continuous, categorical).
3. Variable values
   - Variable values refer to the information contained in each column. Every variable has pre-determined allowable values.
   - Examples of setting allowable values for different types of variables include:

    - Categorical character variable: "yes" | "no"
    - Numeric variable: *1–25*
    - Date variable: *2023-08-01 to 2023-12-15*
    - Free text character variable: Any value is allowed
- Anything outside of your expected values or ranges is considered an error.

4. Variable labels
    - A variable label is the human readable description of what a variable represents.
    - This may be a label that you, as the variable creator, assigns (e.g., "Treatment condition") or it may be the actual wording of an item (e.g., "Do you enjoy pizza?").

3.1.2 Rows

The rows in your dataset are aligned with subjects (also called records or cases) in your data. Subjects in your dataset may be students, teachers, schools, locations, and so forth. The unique subject identifier variable mentioned in Section 3.1.1, will denote which row belongs to which subject.

3.1.3 Cells

The cells are the observations associated with each case in your data. Cells are made up of key/value pairs, created at the intersection of a column and a row (see Figure 3.2). Consider an example where we collect a survey (also called a questionnaire) from students. In this dataset, each row is made up of

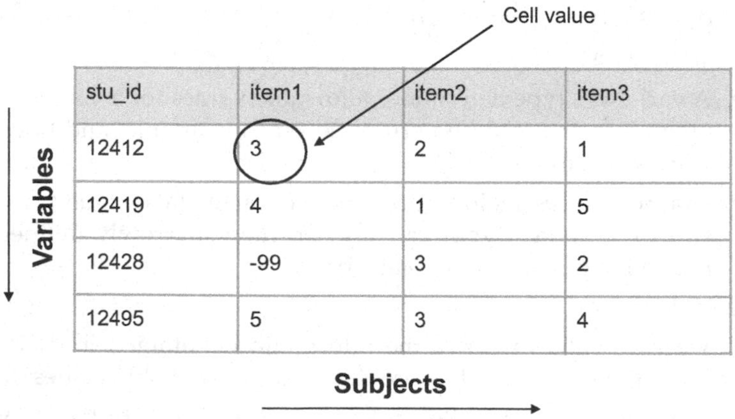

Cell value

Variables

stu_id	item1	item2	item3
12412	3	2	1
12419	4	1	5
12428	-99	3	2
12495	5	3	4

Subjects

FIGURE 3.2
Representation of a cell value.

a unique student in our study, each column is an item from the survey, and each cell contains a value that corresponds to that row/column pair (i.e., that participant and that question).

3.2 Dataset Organization Rules

In order for your dataset to be machine-readable and analyzable, it should adhere to a set of organizational rules (Broman and Woo 2018; Wickham 2014).

1. The first rule is that data should make a rectangle (Figure 3.3). The first row of your data should be your variable names (only use one row for this). The remaining data should be made up of values in cells.
2. Column values should be consistent (Figure 3.4). Both humans and machines have difficulty categorizing information that is not measured, coded, or formatted consistently.
   - For text categorical values, use controlled vocabularies and keep consistent spelling, case, and spacing.
   - For date values, keep the format consistent.
   - For numeric values, measure in consistent units and keep consistent decimal places.
3. Columns should adhere to your expected variable type (Figure 3.5).
   - For example, if you have a numeric variable, such as age, but you add a cell value that is text, your variable no longer adheres to your variable type. Machines will now read this variable type as character.
4. A variable should only collect one piece of information (Figure 3.6). This allows you to more easily work with your variables.
   - For example, rather than combining the number of incidents and the number of enrolled students in the same variable, separate this information into two variables. This allows you to aggregate information as needed (e.g., calculate an incident rate).
5. All cell values should be explicit (Figure 3.7). This means all cells that are not missing values should be filled with a value.
   - Consider why a cell value is empty
     - If a value is actually missing, you can either leave those cells blank or fill them with your pre-determined missing values (e.g., -99). See Section 9.5.1 for ideas on coding missing values.

- If a cell is left empty because it is implied to be the same value as above, the cells should be filled with the actual data.
- If an empty cell is implied to be *0*, fill the cells with an actual *0*.

6. All variables should be explicit (Figure 3.8). No variables should be implied using color coding.
    - If you want to indicate information, add an indicator variable to do this rather than cell coloring.

Not a rectangle

	12412	12419	12428	12495
age_mth	74	74	76	79
	12412	12419	12428	12495
read_raw	52	38	43	54
	12412	12419	12428	12495
read_ss	75	62	67	76

Rectangle

stu_id	age_mth	read_raw	read_ss
12412	74	52	75
12419	74	38	62
12428	76	43	67
12495	79	54	76

FIGURE 3.3
A comparison of non-rectangular and rectangular data.

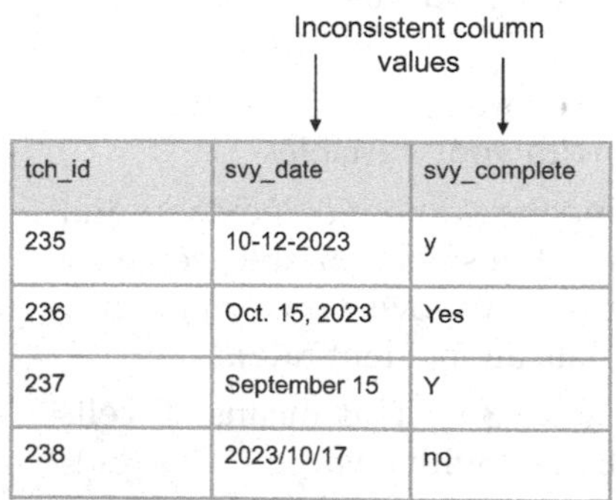

Inconsistent column values

tch_id	svy_date	svy_complete
235	10-12-2023	y
236	Oct. 15, 2023	Yes
237	September 15	Y
238	2023/10/17	no

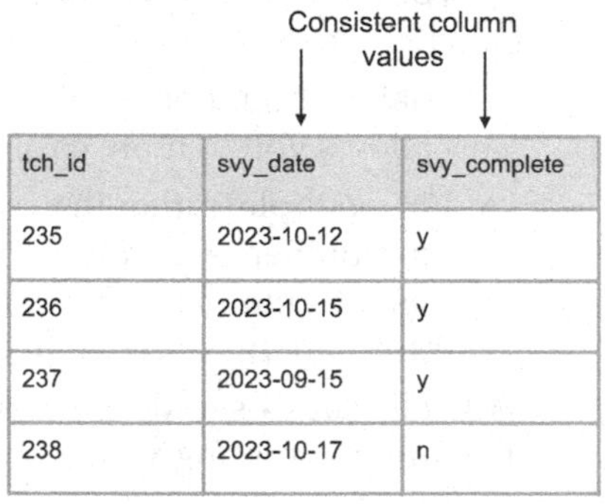

Consistent column values

tch_id	svy_date	svy_complete
235	2023-10-12	y
236	2023-10-15	y
237	2023-09-15	y
238	2023-10-17	n

FIGURE 3.4
A comparison of inconsistent and uniform variable values.

Character variable

tch_id	age_yr
235	22
236	24
237	49 years old
238	36..0

Space before 24 makes this entry text
Text added makes this entry text
Double decimal point makes this entry text

Numeric variable

tch_id	age_yr
235	22
236	24
237	49
238	36

FIGURE 3.5
A comparison of variables adhering and not adhering to a data type.

Two things in one variable

sch_id	level	incident_rate
31	elementary	55/250
35	elementary	72/303
42	middle	140/410
43	high	219/552

Two things in two variables

sch_id	level	incident	enrollment
31	elementary	55	250
35	elementary	72	303
42	middle	140	410
43	high	219	552

FIGURE 3.6
A comparison of two things being measured in one variable and two things being measured across two variables.

Not explicit values

sch_id	year	grade	n_stud
31	2023	3	100
		4	80
		5	90
35	2023	3	98
		4	88
		5	91

Explicit values

sch_id	year	grade	n_stud
31	2023	3	100
31	2023	4	80
31	2023	5	90
35	2023	3	98
35	2023	4	88
35	2023	5	91

FIGURE 3.7
A comparison of variables with empty cells and variables with not empty cells.

Cell color indicates treatment condition

Not explicit variables

stu_id	date	read_raw
12412	2023-04-13	52
12419	2023-04-12	38
12428	2023-04-13	43
12495	2023-04-11	54

Explicit variables

stu_id	date	read_raw	treatment
12412	2023-04-13	52	0
12419	2023-04-12	38	1
12428	2023-04-13	43	1
12495	2023-04-11	54	0

FIGURE 3.8
A comparison of information being indicated through cell color and information being provided in an indicator variable.

3.3 Linking Data

Up until now we have been talking about one, standalone dataset. However, it is more likely that your research project will be made up of multiple datasets, collected from different participants, using a variety of instruments, and possibly across different time points. At some point you will most likely need to link those datasets together.

In order to think about how to link data, we need to discuss two things, database design and data structure.

3.3.1 Database Design

A database is "an organized collection of data stored as multiple datasets" (USGS 2023). Sometimes this database is actually housed in a database software system (such as SQLite or FileMaker), and other times we are loosely using the term database to simply define how we are linking datasets together that are stored individually (i.e., flat files). No matter the storage system, the general concepts here will be applicable.

In database terminology, each dataset we have is considered a "table". Each table includes one or more variables that uniquely define rows in your data (i.e., a primary key). Tables may also contain variables associated with unique values in another table (i.e., foreign keys) (Wickham, Çetinkaya-Rundel, and Grolemund 2023). See Figure 3.9 for an example of three tables that contain primary keys (denoted by rectangles) and foreign keys (denoted by ovals). Furthermore, tables can be joined either horizontally or vertically.

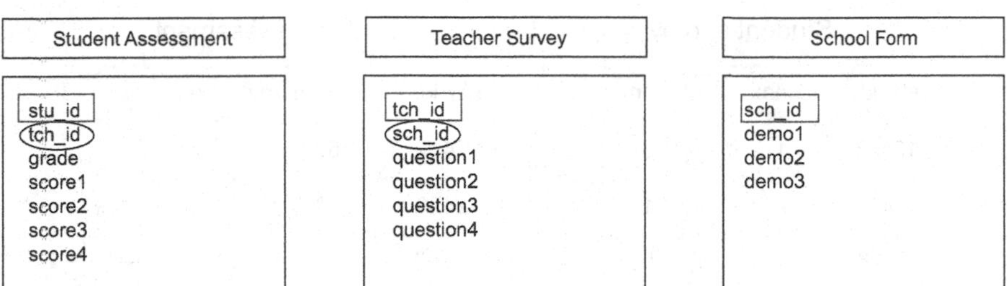

FIGURE 3.9
Three tables with primary and foreign keys.

3.3.1.1 Horizontal Joins

Joining tables horizontally, also called merging, involves matching rows by one (e.g., `stu_id`) or more (e.g., `first_name` and `last_name`) keys, resulting in a table of combined information. With the exception of primary and foreign keys which are often identically named across tables, it is important that all variables are uniquely named across tables when joining horizontally.

There are several different types of horizontal joins (e.g., left, right, inner, full). While diving into the different types of joins is outside of the scope of this book, resources for further learning will be provided at the end of this section. For now, just assume that the joins we discuss are all left joins (i.e., all records from our first table will be matched to any available records in our second table).

To better understand horizontal joins, let's take the simple example in Figure 3.10, where we only have a primary key (`stu_id`) in each table, no foreign keys. Here we collected data from students using two different instruments (a survey and an assessment). When we join these tables on our primary key, it will be a one-to-one merge because each student only appears once in each table.

However, we are often not only collecting data using a variety of instruments, we are also collecting nested data across different entities (e.g., students, nested in classrooms, nested in schools). Let's look at another example where we collected data from both students (an assessment) and teachers (a survey). Figure 3.11 shows how we can now link the foreign key in the student assessment (`tch_id`) with the primary key in the teacher survey (`tch_id`). In this scenario, we are doing a many-to-one join (i.e., multiple students are associated with the same teacher), meaning upon merging, teacher data will be repeated for all students in their classroom.

As you can imagine, as we add more tables, the database structure begins to become even more complex. Figure 3.12 is an example where we collected data from students (a survey and an assessment), from teachers (a survey and an observation), and from schools (an intake form). While the linking structure begins to look more complex, we see that we can still link all of our data through primary and foreign keys. Tables within participant types can be linked by primary keys, and tables across participant types can be linked by foreign keys.

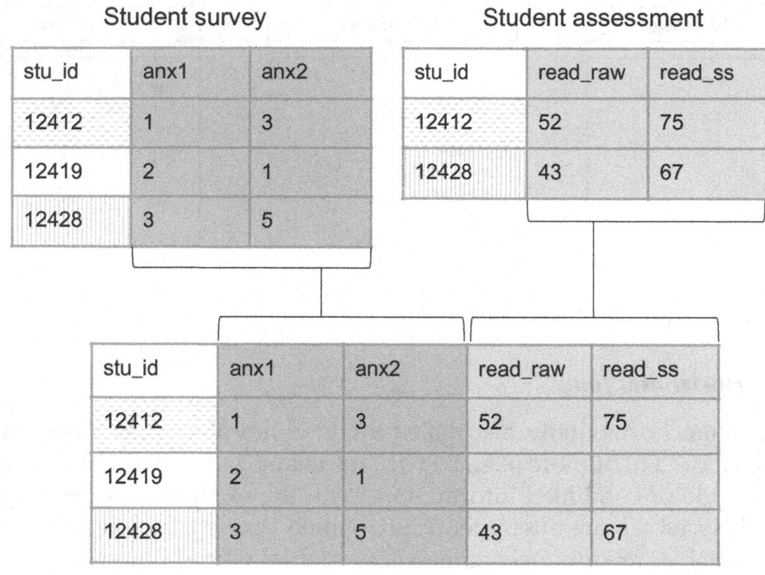

Student survey

stu_id	anx1	anx2
12412	1	3
12419	2	1
12428	3	5

Student assessment

stu_id	read_raw	read_ss
12412	52	75
12428	43	67

stu_id	anx1	anx2	read_raw	read_ss
12412	1	3	52	75
12419	2	1		
12428	3	5	43	67

Merged data

FIGURE 3.10
Linking data through primary keys.

Student assessment

stu_id	tch_id	s_read_raw	s_read_ss
12412	235	52	75
12419	238	38	62
12428	235	43	67
12495	238	54	76

Teacher survey

tch_id	t_exp_yr	t_profdev
235	14	1
238	1	0

stu_id	tch_id	s_read_raw	s_read_ss	t_exp_yr	t_profdev
12412	235	52	75	14	1
12419	238	38	62	1	0
12428	235	43	67	14	1
12495	238	54	76	1	0

Merged data

FIGURE 3.11
Linking data through foreign keys.

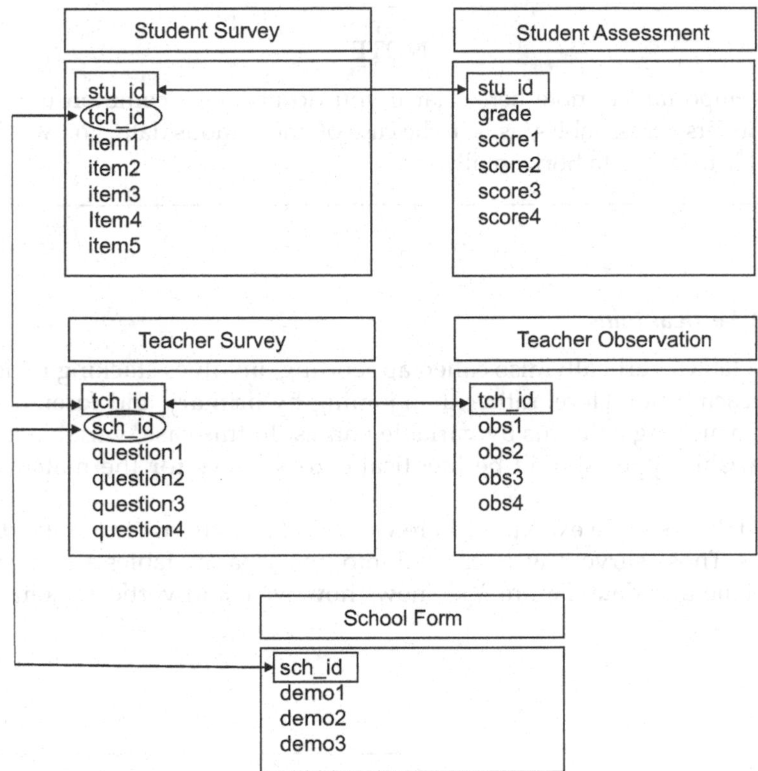

FIGURE 3.12
Linking data through primary and foreign keys.

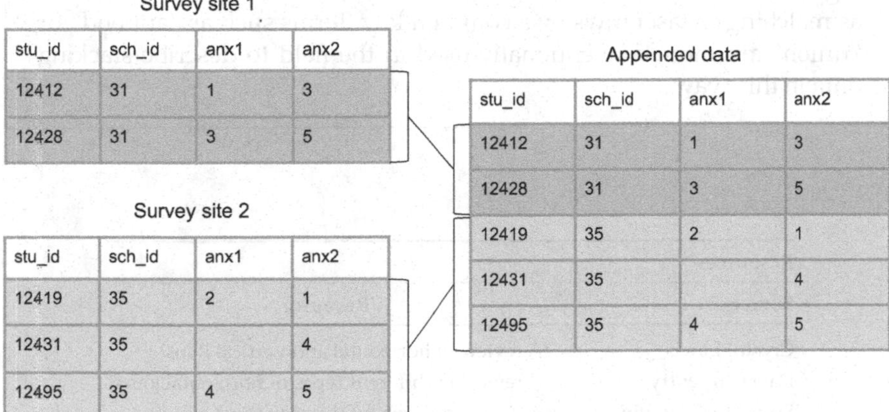

Survey site 1

stu_id	sch_id	anx1	anx2
12412	31	1	3
12428	31	3	5

Survey site 2

stu_id	sch_id	anx1	anx2
12419	35	2	1
12431	35		4
12495	35	4	5

Appended data

stu_id	sch_id	anx1	anx2
12412	31	1	3
12428	31	3	5
12419	35	2	1
12431	35		4
12495	35	4	5

FIGURE 3.13
Appending data across sites.

NOTE

It is important to note here, that if you do not have common unique identifiers across tables, as is in the case of anonymous data, you will be unable to join data horizontally.

3.3.1.2 Vertical Joins

Joining tables vertically, also called appending, involves stacking tables on top of each other. Here, rather than joining by primary and foreign keys, we are matching columns by variable names. In this case, variable names and variable types should be identical across tables for the matching to work.

Let's take a simple example where we collected a survey from two different sites. Those surveys were entered into two separate tables and we want to combine that data. Figure 3.13 shows how we could vertically join those tables.

NOTE

In this Section 3.3 I am loosely using the term "join" to provide a unifying framework for the different ways you can combine data. However, stacking data vertically is not technically considered a join, commonly defined as matching dataset rows by a common key. Terms such as "append" or "union" are more conventionally used in the field to describe stacking data in this way.

Resources

Source	Resource
Crystal Lewis	A review of horizontal and vertical joins[1]
Data Carpentry	A review of different types of horizontal joins[2]
David E. Caughlin	A review of horizontal and vertical joins[3]

3.3.2 Data Structure

When working with longitudinal data, it's important to consider data structure before linking data. Collecting longitudinal, or repeated measures data, typically results in multiple, identically formatted tables, each representing a different wave of data collection. It is common for researchers to join these waves of data for storage or analysis purposes. However, there are two different ways to structure combined longitudinal data—in wide format or long format—and it is important to choose this structure before joining your data.

3.3.2.1 Wide Format

When we structure our data in a wide format, all data collected on a unique subject will be in one row. Subjects should not be duplicated in your data in this format.

To structure data in wide format, we join our tables horizontally. Before joining though, each wave of data collection will be appended to a variable name to create unique names. Figure 3.14 shows of an example of how we could structure two waves of data collection in wide format.

Wave 1 data

stu_id	anx1	anx2
12412	1	3
12419	2	1
12428	3	5

Wave 2 data

stu_id	anx1	anx2
12412	3	4
12419	1	1
12428	2	1

stu_id	w1_anx1	w1_anx2	w2_anx1	w2_anx2
12412	1	3	3	4
12419	2	1	1	1
12428	3	5	2	1

Wide format data

FIGURE 3.14
Example linking tables across time in wide format.

3.3.2.2 Long Format

Another way to structure longitudinal data is in long format. Here a participant can, and often will, repeat in your dataset, and unique rows will now be identified through a combination of variables (e.g., `stu_id` and `wave` together will be your primary key).

To structure data in long format, we join our tables vertically. In this scenario, we no longer add the data collection wave to variable names. However, a time period variable should be added to denote the wave associated with each row of data. Figure 3.15 shows an example of how we could structure two waves of data collection in long format.

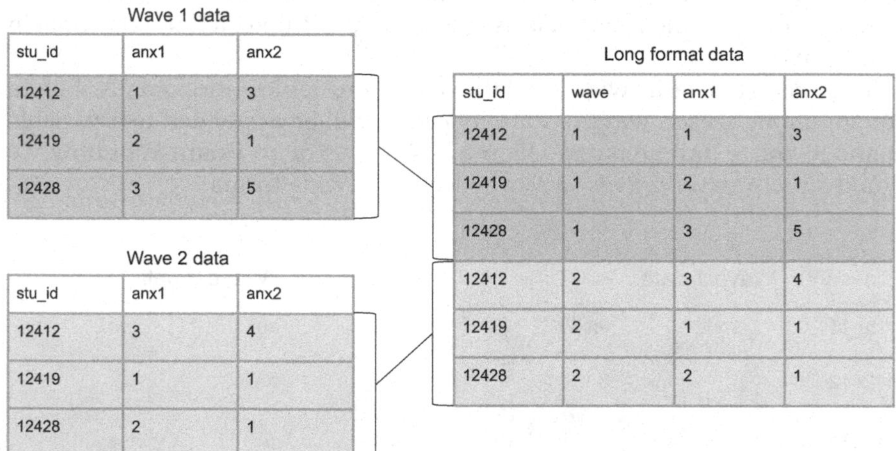

Wave 1 data

stu_id	anx1	anx2
12412	1	3
12419	2	1
12428	3	5

Wave 2 data

stu_id	anx1	anx2
12412	3	4
12419	1	1
12428	2	1

Long format data

stu_id	wave	anx1	anx2
12412	1	1	3
12419	1	2	1
12428	1	3	5
12412	2	3	4
12419	2	1	1
12428	2	2	1

FIGURE 3.15
Example linking tables across time in long format.

3.3.2.3 Choosing Wide versus Long

There are different reasons for structuring your longitudinal data one way or another. Storing data in long format is usually considered to be more efficient than storing in wide format, potentially requiring less memory. However, when it comes time for analysis, specific data structures may be required. For example, repeated measure procedures typically require data to be in wide format, where the unit of analysis is the subject. While mixed model procedures typically require data to be in long format, where the unit of analysis is each measurement for the subject (Grace-Martin 2013). It may be that you structure data in one format for one reason (e.g., storing or sharing), and then restructure data into another format a different reason (e.g., analysis). Luckily, this type of restructuring can be done fairly quickly in many statistical programs.[4] We will further review decision making around data structure in Chapters 14 and 16.

Notes

1 https://cghlewis.com/blog/joins/
2 https://tavareshugo.github.io/r-intro-tidyverse-gapminder/08-joins/index.html
3 https://rforhr.com/join.html
4 https://osf.io/xumg4

4

Human Subjects Data

In addition to understanding how to organize data, we also need a foundational understanding of the types of data we may collect. In the field of education research, we are often working with data that is collected from human subjects. Along with collecting data from people comes the responsibility to secure that data. Data from humans may contain identifiable information increasing the risk that participants can be revealed in a dataset. Human subjects data sometimes also contains information on sensitive topics such as mental health, drug use, or criminal behavior, further increasing risks if participants are identified. Before beginning your project, it is important to assess the type of data you will be collecting and understand the protections that will need to be in place to secure your data. This chapter will briefly review the types of human subjects data you may work with as well as any regulations, organizations, policies, or agreements that may impact how you need to secure your data.

4.1 Identifiability of a Dataset

When working with human subjects there are two types of identifiers you may collect in your study, direct and indirect (see Table 4.1). Direct identifiers are unique to an individual and can be used to identify a participant. Indirect identifiers are not necessarily unique to a particular individual, but if combined with other information they could be used to identify a participant (Kopper, Sautmann, and Turitto 2023a).

A term often used when discussing identifiable information is personally identifiable information (PII). This term broadly refers to information that can be used to identify a participant. There is no agreed-upon list for what fields should be included in a list of PII but generally it includes both the direct and indirect types of information shown in Table 4.1.

DOI: 10.1201/9781032622835-4

TABLE 4.1
Examples of Direct and Indirect Identifiers

Direct Identifiers	Indirect Identifiers
Name	Age
Initials	Race
Address	Ethnicity
Phone number	Income
Email address	Education level
Social security number	Gender
IP Address	Occupation
ID numbers (student ID, state ID)	Date of birth
License numbers	ZIP Code
Account numbers	Special education services
	Data collection date
	Verbatim responses

When collecting data and creating datasets, you will be working with one or more of these four types of data files (UNC Office of Human Research Ethics 2020).

1. Identifiable: Data includes personally identifiable information. It is common for your raw research study data to be identifiable.
2. Coded: In this type of data file, PII has been removed or distorted and names are replaced with a code (i.e., a unique participant identifier). The only way to link the data back to an individual is through that code. The identifying code file (linking key) is stored separate from the research data (see Chapter 10). Coded data is typically the type of file you create after cleaning your raw study data.
3. De-identified: In this type of file, identifying information has been removed or distorted and the data can no longer be reassociated with the underlying individual (the linking key no longer exists). This is typically what you create when publicly sharing your research study data.
4. Anonymous: In an anonymous dataset, no identifying information is ever collected and so there should be little to no risk of identifying a specific participant.

4.2 Data Classification

Data is often classified based on the level of sensitivity (Filip 2023; Macquarie University 2023; University of Michigan, 2023). These levels of sensitivity dictate how the data can be collected, stored, and shared, as well as what the response should be to any data breach. Depending on the institution, the names for these levels, the number of levels, what is included in these levels, and the rules applied to the levels, all vary. While there is variation, here is a general summary of how information may be categorized.

1. Low sensitivity: This data is considered to have no or low risk if disclosed. This typically includes de-identified and anonymous data that does not contain highly sensitive information.
2. Moderate sensitivity: This data is considered to have moderate risk if disclosed, meaning it could adversely affect people. This data may include identifiable information or information that could allow participants to be re-identified within the data itself or using an external source. This data is typically required to be kept confidential by law or other agreements. These data should be protected against unauthorized access.
3. High sensitivity: This data should be under the most stringent security and could cause great harm if disclosed. This data includes PII or information that could allow participants to be re-identified, as well as private or highly sensitive information (e.g., illegal behaviors, medical records) and are typically required to be kept confidential by law or other agreements. These data should be protected against unauthorized access.

It is important to review your institution's data classification levels, or data sensitivity levels, to determine how your specific institution classifies data. These rules may come from an information technology department, an institutional review board (IRB), or a combination of both. Note that different data collection efforts in the same project can be classified in different ways.

4.3 Human Subjects Data Oversight

When working with human subjects data, there are laws, policies, departments, and agreements that may impact how you collect and manage that data. Below we will review some of the most commonly encountered oversight in education research.

4.3.1 Regulations and Laws

1. FERPA: The Family Educational Rights and Privacy Act (FERPA) is a federal law protecting the privacy of student education records. The law applies to elementary and secondary schools, as well as

post-secondary institutions which receive federal funds from the Department of Education. FERPA provides a list of personally identifiable information often contained in education records.[1]

2. HIPAA: The Health Insurance Portability and Accountability Act (HIPAA) provides federal protection for the privacy of protected health information (PHI) collected by covered entities serving patients. The HIPAA Privacy Rule provides a list of 18 identifiers that should be protected.[2]
3. Common Rule: In 1991 the Federal Policy for the Protection of Human Subjects was published, establishing core procedures for human subjects protections. The policy, 45 CFR part 46 (Office for Human Research Protections 2016), included four subparts. Subpart A, known as the "Common Rule", provided a set of protections for human subjects research including informed consent, review by an IRB, and compliance monitoring (National Institute of Justice 2007; Office for Human Research Protections 2009). In 2018 the Common Rule was revised in order to better protect research participants and to reduce administrative burden (Office for Human Research Protections 2018; U.S. Department of Health and Human Services 2018).

4.3.2 Institutions and Departments

1. IRB: An Institutional Review Board (IRB) is a formal organization designated to review and monitor human subjects research and ensure that the welfare, rights, and privacy of participants are maintained throughout the project (Oregon State University 2012). In particular the IRB is concerned with three ethical principles established in the Belmont Report (The National Commission for the Protection of Human Subjects of Biomedical and Behavioral Research 1979); respect for persons (i.e., protecting the autonomy of participants), beneficence (i.e., minimizing harm and maximizing good), and justice (i.e., fair distribution of burdens and benefits)(Duru and Sautmann 2023; Gaddy and Scott 2020). When conducting human subjects research, it is important to review your local IRB's policies and procedures to determine if your study requires IRB approval.
2. IT department: Institutional Information Technology (IT) departments often vet data collection, transfer, and storage tools and are the authority on what tools are approved for research use. They may also be your source for determining classification levels for data security.
3. Office of research or sponsored programs: Institutions often have an administrative body that serves as a signatory authority and can help negotiate terms and conditions for certain types of agreements (Washington University in St. Louis 2023).

4.3.3 External Permission

1. External permission: When planning to conduct research in schools, many districts require researchers to submit requests for research.[3] The requirements for these requests vary by district, but they often include an application or proposal outlining research plans, as well as other supporting documents (e.g., copies of data collection instruments, IRB approval, agreement forms). This submission is then typically reviewed by a committee for possible approval. A similar research or data permission process may also be required when requesting access to non-public data sources such as statewide longitudinal data systems. See Figure 12.6 for an example of what these request processes might look like.

4.3.4 Agreements

1. Informed consent/assent: Often required by an IRB, consent involves informing a participant of what data will be collected for your research study and how it will be handled and used, as well as obtaining a participant's voluntary agreement to participate in your study. If your study involves participants under the age of 18, you may also be required to obtain a participant assent form, in addition to a parent/guardian consent form.
2. DUA: A data use agreement (DUA), also sometimes referred to as a data sharing agreement (DSA), is a contractual agreement that provides the terms and conditions for sharing data. DUAs are commonly written for data sharing when partnering with school districts or state agencies. As an example, a DUA may include the terms for sharing, working with, and storing education records data. However, DUAs can be used to provide guidance for outgoing data as well (i.e., a researcher is sharing their original data with an agency). DUAs can be standalone documents or may be incorporated into other documents such as a memorandum of understanding (MOU).
3. NDA: Non-disclosure agreements (NDAs), sometimes synonymous with confidentiality agreements, restrict the use of proprietary or confidential information (University of Washington 2023) and are legally enforceable agreements. These may be required when partnering with districts or other agencies.

4.3.5 Funders

1. Funding agencies: Along with requiring data management plans, funding agencies may have their own data protection procedures

and may require applicants to submit additional documents agreeing to specific guidelines or outlining their security plans for human subjects data.

4.4 Protecting Human Subjects Data

Throughout the remaining chapters of this book, we will review ways to keep identifiable human subjects data secure in each phase of the research life cycle. With that said, below is a quick review of some of the most important things to remember if you are collecting data that contain PII.

1. In most situations it will be important to get consent to collect identifiers. Consult with your local IRB to determine what is required. See Section 11.2.5 for more information.
2. Collect as few identifiers as possible. Only collect what is necessary. See Section 11.2.1 for more information.
3. Follow rules laid out in applicable laws, policies, and agreements when collecting, storing, and sharing data. This includes, but is not limited to, using approved tools for data collection, capture, and storage, assigning appropriate data access levels, and transmitting data using approved methods. See Chapters 11, 12, 13, and 15 for more information.
4. Remove names in data and replace them with codes (i.e., unique study identifiers). See Sections 10.4 and 14.3.1 for more information.
5. Fully de-identify data before publicly sharing it. See Section 16.2.3.4 for more information.
6. Use data sharing agreements and controlled access as needed when publicly sharing data. See Section 16.2.1 for more information.

Notes

1 https://www.ecfr.gov/current/title-34/subtitle-A/part-99

2 https://www.hhs.gov/hipaa/for-professionals/privacy/special-topics/de-identification/index.html#safeharborguidance

3 Information about these requests can often be found on district websites.

5

Data Management Plan

5.1 History and Purpose

Since 2013, even earlier for the National Science Foundation (NSF), most federal agencies that education researchers work with have required a data management plan (DMP) as part of their funding application (Holdren 2013) (see Figure 5.1). While the focus of these plans is mostly on the future outcome of data sharing, the data management plan is a means of ensuring that researchers will thoughtfully plan a research study that will result in data that can be shared with confidence, and free from errors, uncertainty, or violations of confidentiality. President Obama's May 2013 Executive Order declared that "the default state of new and modernized government information resources shall be open and machine readable" (The White House 2013). In August of 2022, the Office of Science and Technology Policy (OSTP) doubled down on their data sharing policy and issued a memorandum stating that all federal agencies must update their public access policies no later than December 31, 2025, to make federally funded publications and their supporting data accessible to the public with no embargo on their release (Nelson 2022). Even sooner than this, organizations like the National Institutes of Health (NIH) mandated that grant applicants, beginning January 2023, must submit a plan for both managing and sharing project data (National Institutes of Health 2023c). The National Science Foundation also released version 2.0 of their public access plan in February of 2023, describing how the agency plans to ensure that all scientific data, funded by the NSF and associated with peer-reviewed publications, is publicly shared (National Science Foundation 2023).

NOTE

In the last year, agencies have begun revising the phrase "data management plan" to include the word "sharing" to better represent the shifting emphasis on sharing publicly funded data. As an example,

DOI: 10.1201/9781032622835-5

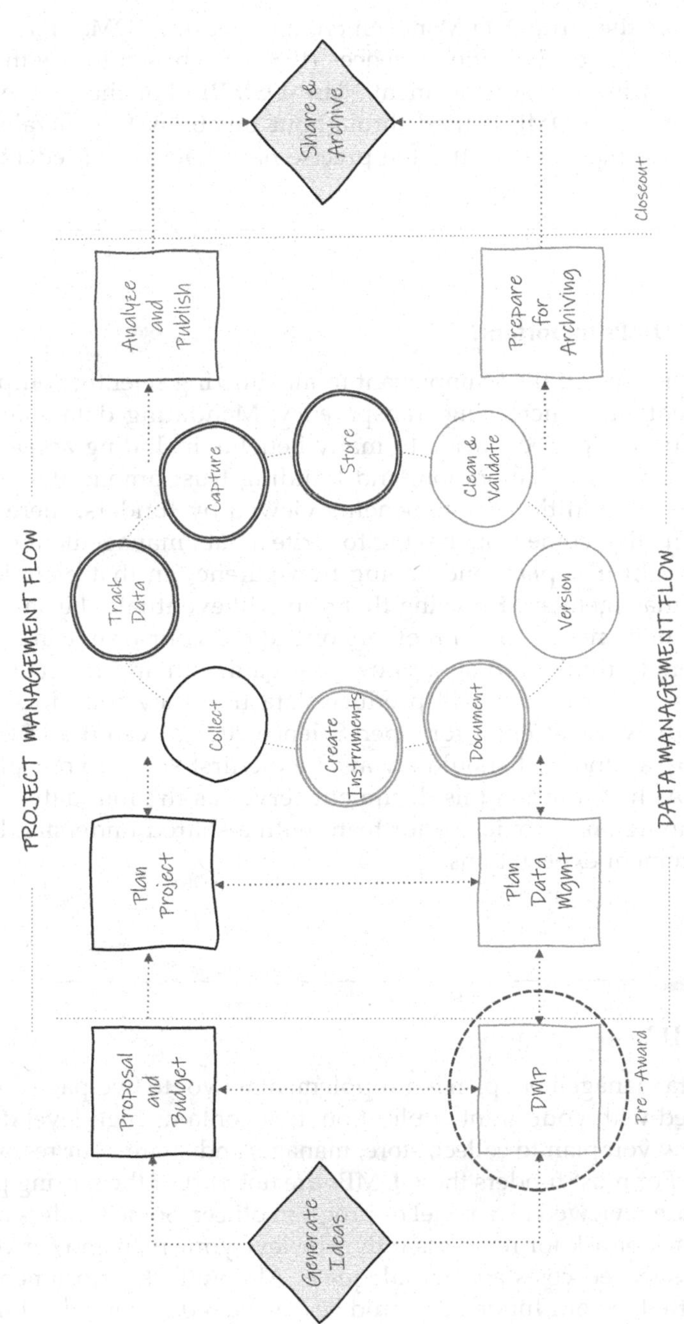

FIGURE 5.1
Data management plan in the research project life cycle.

NIH now uses the term Data Management and Sharing (DMS) Plan,[1] while the Institute of Education Sciences (IES) has chosen to use the term Data Sharing and Management Plan (DSMP).[2] For the sake of simplicity, the term DMP is used throughout this book to generally represent these plans, no matter the precise name, across all federal agencies.

5.1.1 Why Are DMPs Important?

Funding agencies see DMPs as important in maximizing scientific outputs from investments and increasing transparency. Mandating data sharing for federally funded projects leads to many benefits including accelerating discovery, greater collaboration, and building trust among data creators and users. In addition to the benefits viewed by funders, there are intrinsic benefits that come from having to write a data management plan. Having to thoughtfully plan, and having transparency in that plan, lead to better data management. Knowing that you will eventually be sharing your data and documentation with others outside of your team can motivate researchers to think hard about how to organize their data management practices in a way that will produce data that they trust to share with the outside world (Center for Open Science 2023). Even if a DMP is not required by a funder, it should always be the first step of your planning process. Although brief, this document serves as the foundation for all future planning and provides your team with a shared understanding of data management expectations.

5.2 What Is It?

Typically, a data management plan is a supplemental two- to five-page document, submitted with your grant application, that contains high-level decisions about how you plan to collect, store, manage, and share your research data products. For most funders these DMPs are not part of the scoring process, but they are reviewed by a panel or program officer. Some funders may provide feedback or ask for revisions if they believe your plan and/or your budget and associated costs are not adequate. Although this document is usually submitted to your funder, it should be considered a living document to be updated as plans change throughout a study.

5.2.1 What to Include?

What to include in a DMP varies some across funding agencies and the landscape of requirements is currently evolving. You should check each funding agency's site for their specific DMP requirements when submitting a proposal. With that said, there are generally ten common categories covered in a data management plan (Center for Open Science 2023; Gonzales, Carson, and Holmes 2022; ICPSR 2020; Michener 2015) which we will review below.

1. Description of data to be shared (see Chapters 11, 12, 14, and 16)
    - What is the source of data? (e.g., surveys, assessments, observations, extant data)
    - How will data be cleaned and curated prior to data sharing?
    - What will the level of aggregation be? (e.g., item-level, summary data, metadata only)
        - Datasets from a project may need to be shared in different ways due to legal, ethical, or technical reasons.
    - Will both raw and clean data be shared?
    - What is the expected number of files? Expected number of rows/cases in each file?
2. Format of data to be shared (see Chapters 14 and 16)
    - Will data be in an electronic format?
    - Will it be provided in a non-proprietary format? (e.g., CSV)
    - Will more than one format be provided? (e.g., SPSS and CSV)
    - Are there any tools needed to manipulate or reproduce shared data? (e.g., software, code)
        - Provide details for those tools. (e.g., how they can be accessed, version number, required operating system)
3. Documentation to be shared (see Chapters 8 and 16)
    - What documentation will you share?
        - Consider project-level, dataset-level, and variable-level documentation.
    - What format will your documentation be in? (e.g., XML, CSV, PDF)
4. Standards (see Chapters 8, 11, and 16)
    - Do you plan to use any standards for things such as metadata, data collection, or data formatting?
5. Data preservation (see Chapter 16)
    - Where will data be archived for public sharing?

- Many agencies are now requiring applicants to name a specific data repository in this section.
- What are the desirable characteristics of the repository?[3] (e.g., unique persistent identifiers assigned to data, metadata collected, records provenance, licensing options)
- When will you deposit your study data in the repository and for how long will data remain accessible?
- How will you enable discoverability and reuse of data?

6. Access, distribution, or reuse considerations (see Chapters 4 and 16)
    - Are there any legal, technical, or ethical factors affecting reuse, access, or distribution of your data?
    - Will any data be restricted?
    - Are access controls required (e.g., a data use agreement, data enclave)?
7. Protection of privacy and confidentiality (see Chapters 4, 14, and 16)
    - Do participants sign informed consent agreements? Does the consent communicate how participant data are expected to be used and shared?
    - How will you prevent disclosure of personally identifiable information when you share data?
8. Data security (see Chapter 13)
    - How will security and integrity of data be maintained during a project? (e.g., consider data storage, access, backup, and transfer)
9. Roles and responsibilities (see Chapter 7)
    - What are the staff roles in management and preservation of data?
    - Who ensures accessibility, reliability, and quality of data?
    - Is there a plan if a core team member leaves the project or institution?
10. Preregistration
    - Where and when will you preregister your study?

Again, the specifics of what should be included in each category will vary by funder. Here are sites to visit to learn more about the DMP requirements for four common federal education research funding agencies.

- Institute of Education Sciences[4,5]
- National Institutes of Health[6]
- National Institute of Justice[7]
- National Science Foundation[8]

5.3 Creating a Data Sources Catalog

In preparation for writing your DMP, it can be helpful to create a data sources catalog that allows you to visually see what data you are collecting, what the sensitivity level of each source is, and how data will be collected, managed, stored, and shared (Filip 2023). This type of catalog cannot only help you write your DMP but can also serve as an excellent planning or discussion tool throughout your entire project.

In setting up this rectangular formatted document, each row represents a unique data collection effort, and each column (or field) represents information about that effort. Some fields you can add to this catalog include:

- Source information
  - Instrument (e.g., survey, assessment)
  - Record level (i.e., who is this instrument collected on)
  - Who completes the instrument (e.g., rater, participant)
  - Measures included in the instrument
- Collection and capture method
- Data collection waves (i.e., how often will you collect this data source)
- Planned number and size of data files for each source (e.g., two student assessment files (T1, T2), with ~500 rows per file)
- PII included
- Sensitivity level based on your institution's policies (consider levels both before and after de-identification)
- Data storage and access plan
- Data ownership
- How confidentiality will be secured
- Data sharing method

Figure 5.2 is a simplified example of building this catalog for a hypothetical study. Ultimately, each data source in your catalog, multiplied by the number of cohorts and/or waves it is collected, will give you an estimate of the final number of distinct data files at the end of your study. In Figure 5.2, if we only collected data for one year, we would end up with six datasets at the end of our study, one teacher-level file and five student-level files. In Chapter 16, we will discuss whether to share these as separate datasets, or larger files combined by unit of analysis (e.g., combined student-level file, combined teacher-level file).

Instrument	Measures	Collection/Capture Method	Time Periods Collected	Direct or Indirect PII	Sensitivity Level	Data Storage and Access	Secure Confidentiality	Data Sharing Method
Teacher self-report survey	Teaching experience, Professional development, Math anxiety	Collected in Qualtrics, Exported to SPSS for cleaning	T1	Name, Open-ended responses	Moderate	Stored on institution network drive, Access limited to need-to-know personnel	Consent, Coded data, PII removed, Open-ended responses categorized	Clean only, Item-level, CSV format, OSF repository open access
Student assessment	Math achievement test	Collected on paper, Raw values entered into scoring program, Scores exported to CSV for cleaning	T1, T2	Name, DOB	Moderate	Paper stored in locked filed cabinets, Electronic files stored on institution network drive, Access limited to need-to-know personnel	Consent, Coded data, PII removed	Clean only, Summary scores, CSV format, OSF repository open access
Teacher rating of student survey	Math confidence, Math anxiety	Collected in Qualtrics, Exported to SPSS for cleaning	T1, T2	Name	Moderate	Stored on institution network drive, Access limited to need-to-know personnel	Consent, Coded data, PII removed	Clean only, Item-level, CSV format, OSF repository open access
Student school records	Demographics, Attendance, Discipline, State testing scores	CSV file received from districts, Further cleaning performed as needed	T2	Name, DOB, Student ID Race, Gender	Moderate	Stored on institution network drive, stored according to partner DUA, Access limited to need-to-know personnel	Consent, Coded data, PII removed, Collapse small demographic categories as needed	Clean only, Metadata shared in OSF repository, DSA required to access item-level data in CSV format

*T1 = Oct; T2 = May

FIGURE 5.2
Example data sources catalog.

5.4 Getting Help

Since DMPs are written before a project is funded, and therefore before additional staff members may be hired, oftentimes the investigators developing the grant proposal are the ones who write the DMP. However, when constructing your DMP it is well worth your time to enlist help. If you have an existing data manager or data team, you will most certainly want to consult with them when writing your plan to ensure your decisions are feasible. If you work for a university system, your research data librarians are also excellent resources with a wealth of knowledge about writing comprehensive data management plans. Also, if you plan to share your final data in a repository or institutional archive you will want to contact their team when writing your plan as well. The repository may have its own requirements for how and when data must be shared, and it is helpful to outline those guidelines in your data management plan at the time of submission. Last, you may want to obtain the help of your colleagues. Your colleagues have likely written DMPs before and many people are willing to share their plans as a way to help others better understand what to include.

As mentioned before, your DMP is a living document, and you can always update your plan during or after your project completion. It may be helpful to keep in contact with your program officer regarding any potential changes throughout your project.

If you are looking for guidance in writing a DMP, a variety of generic DMP templates for different federal agencies are available, as well as actual copies of submitted DMPs that some researchers graciously make publicly available for example purposes. Furthermore, the DMPTool (https://dmptool.org), a free, open-source, online application, allows users to create and share data management plans using pre-defined funding agency templates.

Templates and Resources

Source	Resource
DMPTool	Templates organized by funding agencies[9]
Figshare	DMP prompts specific to depositing data with Figshare[10]
Hao Ye, et al.	NIH DMS Plan checklist[11]
Harvard Longwood Medical Area RDM Working Group	Annotated DMP template[12]
ICPSR	NIH DMS Plan template with specific recommendations for depositing data with ICPSR[13]
NIH	Sample DMS Plan for human survey data[14]
Sara Hart	A submitted DMP that is publicly available for example purposes[15]
UMN Libraries	Submitted DMP examples from University of Minnesota researchers[16]

5.5 Budgeting

Effective data management requires a significant investment. Funding agencies acknowledge that there are costs associated with implementing your data management plan and allow you to explain these costs in your budget narrative. Costs associated with the entire data life cycle should be considered (Cruse 2011), and should include things such as personnel expenses, as well as fees for tools and services. Make sure to review your funder's documentation for information about allowable costs and time frame for incurring costs. Allowable costs might include things such as the following (National Institutes of Health 2023a; UK Data Service 2023):

- Infrastructure or tools required to organize, document, or store data
- Curating and de-identifying data
- Developing data documentation
- Depositing data for long-term sharing in a repository

It can be difficult to estimate the costs of everything that is associated with the vast landscape of managing data. The necessary dollar amount will vary depending on the size of your project, the expertise needed, and your specific data management plan. Recommendations suggest allocating anywhere from 5% to 30% of your budget for data stewardship (Mons 2020; Mohr et al. 2024; Reynolds et al. 2014). Luckily a few organizations have developed resources to aid in estimating these costs. Exercise caution when using tools though; they may not always account for every expense and could result in an underestimation of costs (Michigan State University 2023).

Resources

Source	Resource
National Institutes of Mental Health Data Archive	NDA Data Submission Cost Estimation Tool[17]
UK Data Service	Data management costing tool and checklist[18]
University of Twente	Estimating RDM costs review list[19]
Utrecht University	Estimating the costs of data management review list[20]

Notes

1 https://grants.nih.gov/grants/guide/notice-files/NOT-OD-21-013.html
2 https://ies.ed.gov/funding/pdf/2024_84305a.pdf

3 https://repository.si.edu/bitstream/handle/10088/113528/Desirable%20Characteristics%20of%20Data%20Repositories.pdf
4 https://ies.ed.gov/funding/datasharing_implementation.asp
5 https://ies.ed.gov/funding/pdf/2024_84305a.pdf
6 https://sharing.nih.gov/data-management-and-sharing-policy/planning-and-budgeting-DMS/writing-a-data-management-and-sharing-plan
7 https://nij.ojp.gov/funding/data-archiving
8 https://new.nsf.gov/funding/data-management-plan
9 https://dmptool.org/public_templates
10 https://help.figshare.com/article/how-to-write-a-data-management-plan-dmp-and-include-figshare-in-your-data-sharing-plans
11 https://osf.io/awypt/
12 https://osf.io/ztjf2
13 https://www.icpsr.umich.edu/files/ICPSR/nih/ICPSR-NIH-DMS-Plan-Template_2023.docx
14 https://www.nichd.nih.gov/sites/default/files/inline-files/Example_DMS_Plan-Human-Survey-NIH_Format_Page_V2.pdf
15 https://figshare.com/articles/preprint/Example_of_a_Data_Management_Plan/13218743
16 https://www.lib.umn.edu/services/data/dmp-examples
17 https://s3.amazonaws.com/nda.nih.gov/Documents/NDA_Data_Submission_Cost_Estimation_Tool.xlsx
18 https://ukdataservice.ac.uk//app/uploads/costingtool.pdf
19 https://www.utwente.nl/en/service-portal/services/lisa/resources/files/library-public/dcc-rdm-costs-estimation.pdf
20 https://www.uu.nl/en/research/research-data-management/guides/costs-of-data-management

6

Planning Data Management

Planning data management is distinct from the two- to five-page data management plan (DMP) discussed in Chapter 5. Here we are spending a few weeks, maybe months, meeting regularly with our team and gathering information to develop detailed instructions for how we plan to manage data according to our DMP. This data management planning happens at the same time that the project team is planning for project implementation (e.g., how to collect data, how to hire staff, what supplies are needed, how to recruit participants, how to communicate with sites) (see Figure 6.1). Team members such as investigators, project coordinators, and data managers may be assisting in both planning processes.

6.1 Why Spend Time on Planning?

Funder required data management plans are hopeful outlines for future practices. However, the broad theory behind our DMPs does not actually prepare us for the complex implementation of those plans in practice (Borycz 2021). Therefore, it is important to spend time, before your project begins, planning and preparing for data management. It is an upfront time investment, but this sort of slow science leads to better data outcomes. Reproducibility begins in the planning phase. Taking time to create, document, and train staff on data management standards before your project begins helps to ensure that your processes are implemented with fidelity and can be replicated consistently throughout the entire study.

Planning the day-to-day management of your project data has many other benefits as well. It allows you to anticipate and overcome barriers to managing your data, such as communication issues, training needs, or potential tool issues. This type of planning also saves you time in the long run, removing the last-minute scrambling that can occur when trying to organize your data at the end of a project. Last, this type of planning can mitigate errors. Viewing errors as problems created by poorly planned workflows, rather than individual failures, helps us to see how data management planning can lead to better data

DOI: 10.1201/9781032622835-6

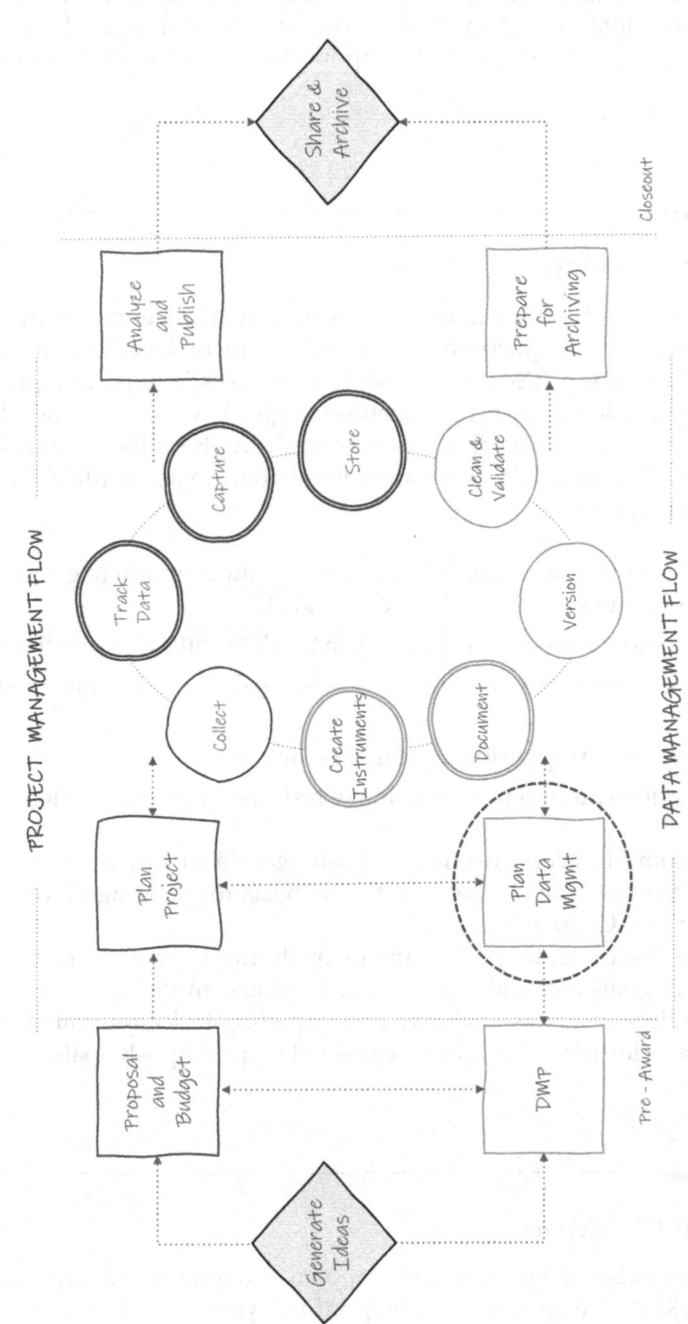

* Double outline means both teams typically collaborate on this phase

FIGURE 6.1
Planning in the research project life cycle.

(Strand 2021). While data management planning cannot remove all chances of errors creeping into your data (Eaker 2016), it can most certainly reduce those errors and prevent them from "compounding over time" (Alston and Rick 2021, 4).

6.2 Goals of Planning

This planning phase should include a series of regular meetings with core decision-makers. For this purpose, it can be helpful to form a data management working group (DMWG), consisting of investigators, key project staff, and other decision-makers (e.g., methodologists), who can attend planning meetings and provide feedback as needed throughout the project (Van Bochove, Alper, and Gu 2023). There are several goals to accomplish during these planning meetings:

1. Further flesh out project goals laid out in a grant proposal (e.g., confirm measures being collected in your study).
2. Finalize a timeline for goals (e.g., confirm the data collection timeline).
3. Lay out specific tasks required to accomplish data management plans.
4. Assign roles and responsibilities for specific tasks.
5. Make decisions around how to manage tasks and communication.

Make sure to come to every meeting with an agenda to stay on track and to take detailed notes. These notes will be the basis for creating all of your documentation (see Chapter 8).

At the end of the planning period, the team should have a clear plan for what the project goals are, when goals should be accomplished, how goals will be accomplished, who is in charge of completing tasks associated with goals, and what additional resources are needed to accomplish goals.

6.3 Planning Checklists

Along with your existing data management plan and other grant application materials, checklists are great tools to help inform your meeting agendas as you work through this planning process with your team. Below are sample checklists, one for each phase of the research cycle. These checklists can be added to or amended and brought to your planning meetings to help your team think through the various data management decisions that need to be made at each phase of your research project.

Planning checklists

- Roles and Responsibilities (https://osf.io/ghtc6)
- Task Management (https://osf.io/8x3s4)
- Documentation (https://osf.io/bckh8)
- Data Collection (https://osf.io/ckwtr)
- Data Tracking (https://osf.io/ahz4f)
- Data Capture (https://osf.io/pc8nt)
- Data Storage and Security (https://osf.io/y3z9s)
- Data Cleaning (https://osf.io/7t4rg)
- Data Archiving (https://osf.io/u4exh)
- Data Sharing (https://osf.io/jkgwz)

NOTE

If this is your first time working through this book, these checklists are a great way to summarize content from each chapter. As you learn best practices for a phase, pull up the checklist specific to that chapter to begin thinking through which practices are feasible for your specific project.

6.3.1 Decision-Making Process

This decision-making process during this planning phase is personalized. Borghi and Van Gulick (2022) view this process as a series of steps that a research team chooses, out of all the many possibilities not chosen. Maybe you won't always be able to implement the best practices, but you can decide what is good enough for your team based on motivations, incentives, needs, resources, skill set, and requirements.

For example, one team may choose to collect survey data on paper because their participants are young children, hand-enter it into Microsoft Excel because that is the only tool they have access to, and double-enter 20% because they don't have the capacity to enter more than that. Another team may choose to collect paper data because they are collecting data in the field, hand-enter the data into FileMaker because that is the tool their team

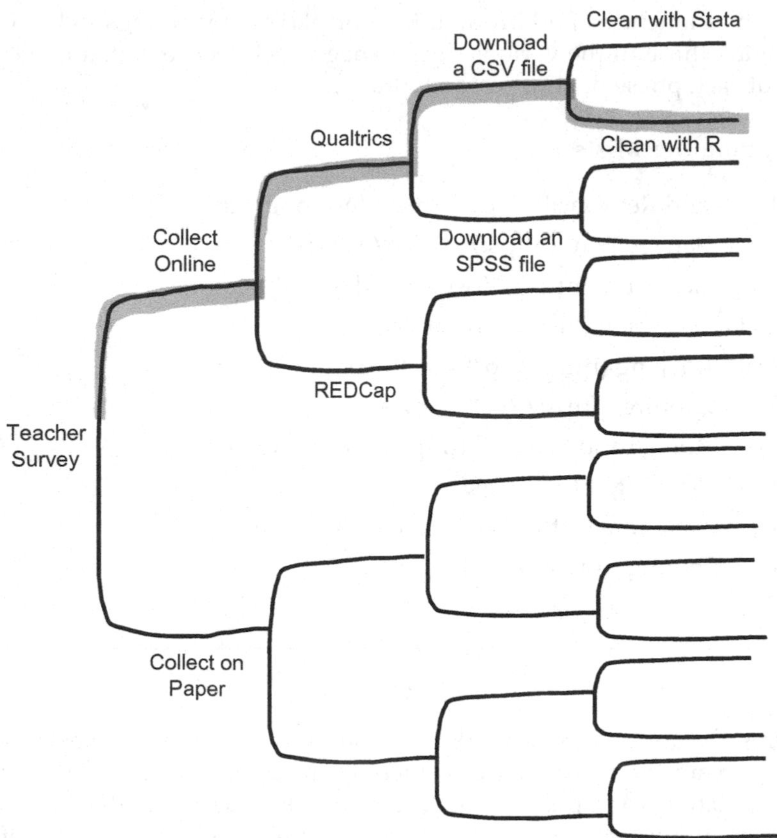

FIGURE 6.2
A simplified decision-making process.

is familiar with, and double-enter 100% because they have the budget and capacity to do that.

Figure 6.2 is a very simplified example of the decision-making process, based on a Borghi and Van Gulick (2022) flow chart. Of course, in real life we are often choosing between many more than just two options!

6.3.2 Checklist Considerations

It's important to consider how each team and project are unique as you work through these planning checklists. A technique that might work well for one team may not work well for another. Make sure to consider the following:

1. All external requirements
   - Do your practices align with the plan laid out in your DMP? If no, you may need to revise your DMP to match your new decisions (remember your DMP is a living document).
   - Do your practices meet all other compliance requirements (e.g., IRB requirements, IT department requirements, consent agreements, data sharing agreements)?
2. The skill set of your team
   - How does the skill set of your team align with the practices you plan to implement? Will additional training be required?
3. Your available tools
   - What tools are available to your team?
   - Are your tools appropriate for the sensitivity level of your files?
   - What is the complexity of your tools? Will additional training be needed?
4. Your budget
   - Do you have the budget to implement all of the practices you want to implement, or will you need to plan something more feasible?
5. Complexity of your project
   - The size of your project, the amount and types of data you are collecting, the number of participants or the populations you are collecting data from, the sensitivity level of the data you are collecting, the number of sites you are collecting data at, and the number of partners and decision-makers you are working with, all factor into your data management planning.
6. Shared investment
   - Is your entire team invested in quality data management?
   - Is the entire team motivated to adhere to the standards and instructions laid out in your data management planning? If no, what safeguards can you implement to help prevent errors from creeping into your data?

6.4 Data Management Workflow

The last step of this planning phase is to build your workflows. Workflows allow data management to be seamlessly integrated into your data collection process. Often illustrated with a flow diagram, a workflow is a repeatable series of sequential tasks (Lucidchart 2019) that help you move through the stages of the research life cycle in an organized and efficient manner. As you walk through your checklists, you can begin to enter your decisions into a workflow diagram that shows actionable steps in your data management process. The order of steps should follow the general order of the data management life cycle (specifically the data collection cycle). Your team will want to have a workflow diagram for every source in your data sources catalog (see Section 5.3). So, for example, if you collect the following three instruments, you will have three workflow diagrams.

- Student online survey
- Student paper assessment
- Student school records

Your diagrams should include the who, what, how, where, and when of each task in the process. Adding these details is what makes the process actionable (Borycz 2021). Your diagram can be displayed in any format that works for you and it can be as simple or as detailed as you want it to be. A template like the one in Figure 6.3 works well for thinking through high-level workflows. Remember, this is a repeatable process. So, while this diagram is linear (steps laid out in the chronological order in which we expect them to happen), this process will be repeated every time we collect this same piece of data.

Figure 6.4 is how we might complete a simplified diagram for a student survey.

But the format truly does not matter. Figure 6.5 is a diagram of the same student survey workflow as Figure 6.4, but with more detail added, and this time using a swimlane diagram where each lane displays the tasks associated with the individual and the iterative processes that occur within and across lanes.

If you have a working data collection timeline (see Section 8.2.6) already created, you can even build time into your workflow. Figure 6.6 is another example of the same survey workflow again, this time displayed using a Gantt chart (Duru, Kopper, and Turitto 2021) in order to better capture the expected timeline.

While these workflow diagrams are excellent for high-level views of what the process will be, we can see that we are unable to put fine details into this visual. So, the last step of creating a workflow is to put all tasks (and all final decisions associated with those tasks) into a standard operating procedure (SOP). In your SOP you will add all necessary details of the process. You can also attach your diagram as an addendum or link your SOPs and diagrams in other ways for reference. We will talk more about creating SOPs in Section 8.2.7.

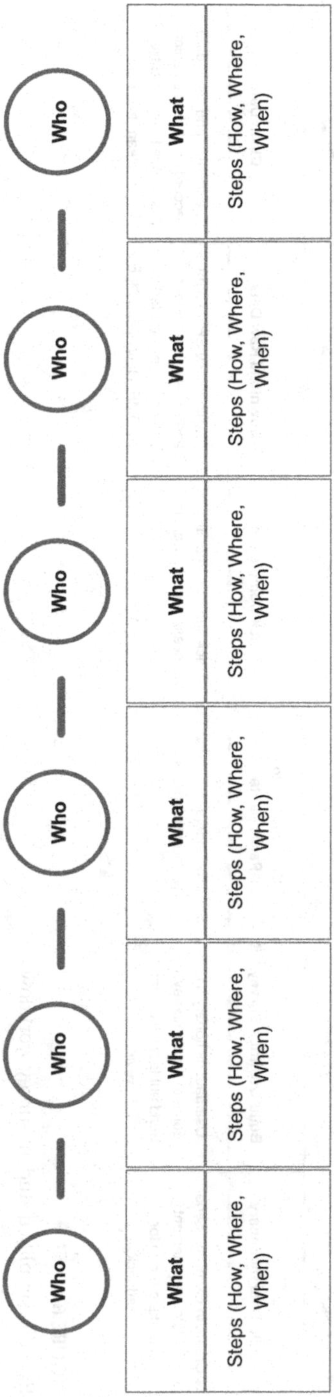

Who	Who	Who	Who	Who	Who
What	**What**	**What**	**What**	**What**	**What**
Steps (How, Where, When)	Steps (How, Where, When)	Steps (How, Where, When)	Steps (How, Where, When)	Steps (How, Where, When)	Steps (How, Where, When)

FIGURE 6.3
A simple workflow template.

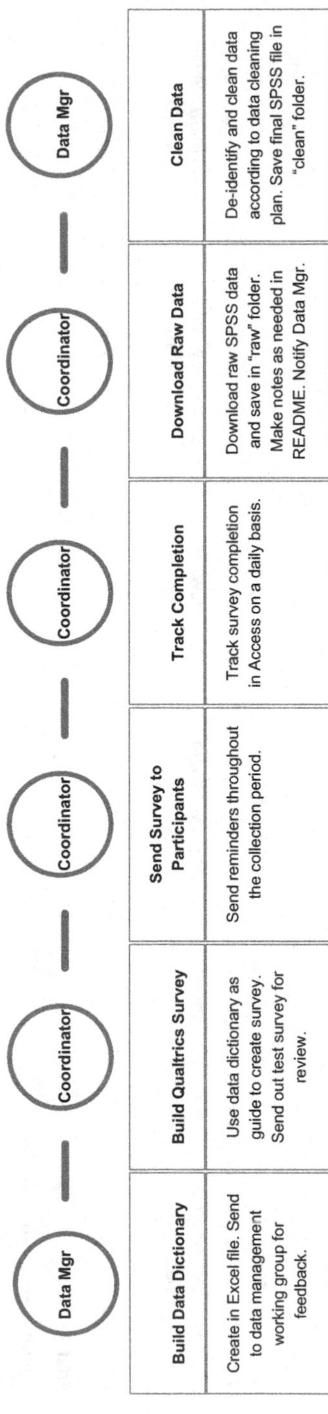

FIGURE 6.4
Example simplified student survey workflow.

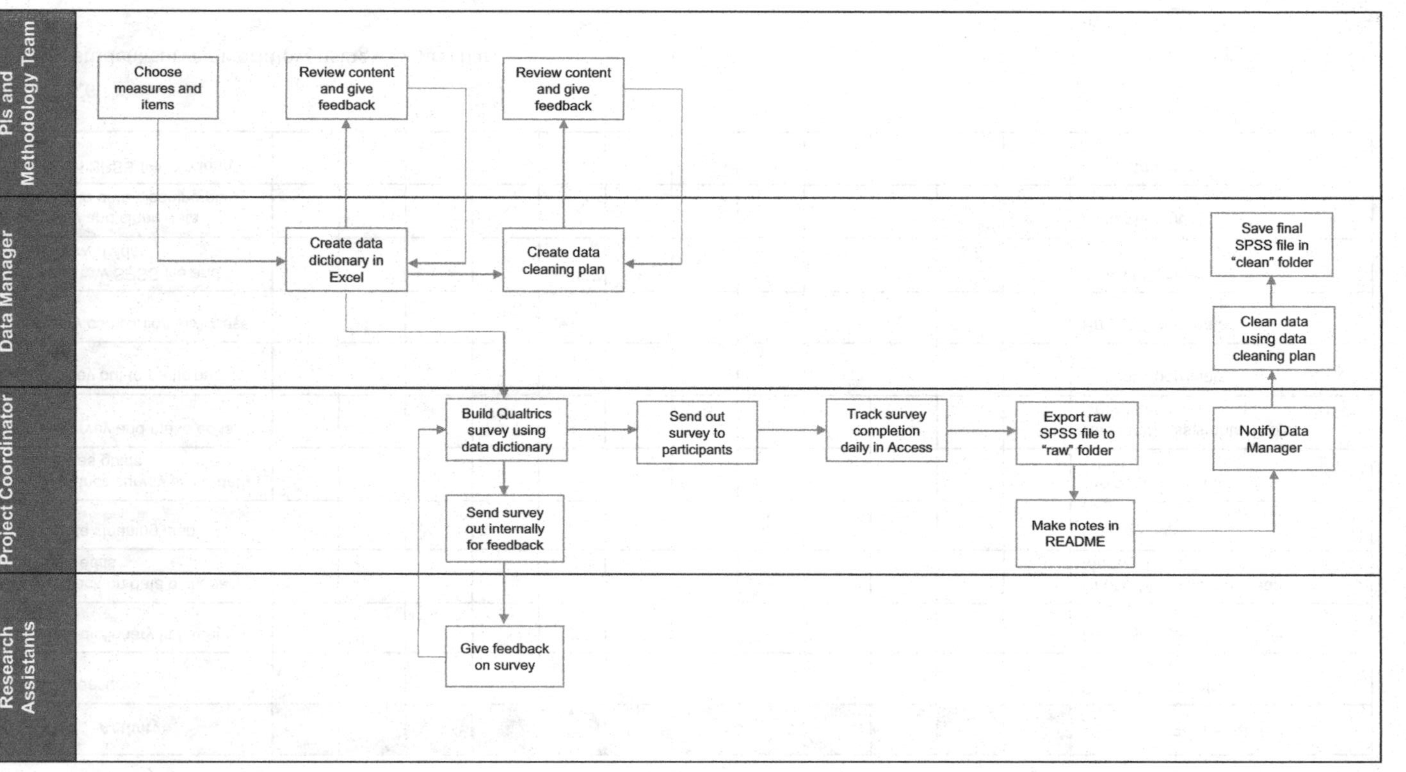

FIGURE 6.5
Example student survey workflow using a swimlane diagram.

Year 1													
Activity	**Aug**	**Sep**	**Oct**	**Nov**	**Dec**	**Jan**	**Feb**	**Mar**	**Apr**	**May**	**June**	**July**	**Responsible**
Choose measures													PI and Methodology Team
Build data dictionary in Excel													Data Manager
Get feedback on data dictionary and make edits													PI and Methodology Team, Data Manager
Create data cleaning plan													Data Manager
Build Qualtrics survey using data dictionary as guide													Project Coordinator
Review survey and make edits													Research Assistants, Project Coordinator
Send survey out to participants													Project Coordinator
Track daily completion in Access													Project Coordinator
Download raw SPSS file and save in "raw" folder													Project Coordinator
De-identify and clean data according to data cleaning plan													Data Manager
Save clean SPSS file in "clean" folder													Data Manager

FIGURE 6.6
Example student survey workflow using a Gantt chart.

6.4.1 Benefits to Visualizing a Workflow

Visualizing your decisions in diagram format has many benefits. First, it allows your team to conceptualize their specific tasks in the process, the timing at which their tasks occur, and any dependencies associated with those tasks. It also allows your team to see how their roles and responsibilities fit into the larger research process (Briney, Coates, and Goben 2020). Showing how data management is integrated into the larger research workflow can help team members view data management as part of their daily routine, rather than "extra work" (Borghi and Van Gulick 2022). And last, reviewing workflows as a team and allowing members to provide feedback may help create buy-in for data management processes, potentially leading to better adherence to practices.

6.4.2 Workflow Considerations

Similar to the questions you need to consider when reviewing your planning checklists, you also need to evaluate the following things when developing your personalized workflows (Hansen 2017).

- Does your flow preserve the integrity of your data? Is there any point where you might lose or comprise data?
- Is there any point in the flow where data is not being handled securely? Can someone gain access to identifiable information they should not have access to?
- Is your flow in accordance with all of your compliance requirements?
- Is your flow feasible for your team (based on size, skill level, motivation, etc.)?
- Is your flow feasible for your budget and available resources?
- Is your flow feasible for the amount and types of data you are collecting?
- Are there any bottlenecks in the workflow? Any areas where resources or training are needed? Any areas where tasks should be re-directed?

6.5 Task Management Systems

While tools such as our checklists, workflow diagrams, and SOPs allow us to document and share our processes, it can be tricky to manage the day-to-day implementation of those processes. The planning phase is a great time to choose a task management system (Gentzkow and Shapiro 2014). Keeping track of various deadlines and communications across scattered sources can be overwhelming and using a task management system may help remove ambiguity about the status of task progress. Rather than having to regularly check in via email for status updates or reading through various meeting notes to learn about decisions made, a task management system allows you to assign tasks to responsible parties, set deadlines based on timelines, track progress, and capture communication and decisions all in one location.

There are many existing tools that allow teams to assign and track tasks, schedule meetings, track project timelines, and document communication. Without endorsing any particular product, some project/task management tools that I know education research teams have used include:

- Trello
- Smartsheet
- Todoist
- Microsoft Planner
- Notion
- Basecamp
- Confluence
- Asana

Of course, as with all processes we've discussed so far, a task management system is only useful if your team is trained to use it, is invested in using it, and actually uses it as part of their daily routine. So, make sure to consider this as you choose what tool, if any, is right for you.

7

Project Roles and Responsibilities

Part of the DMP and planning data management phase, as noted in previous chapters, will include assigning roles and responsibilities (see Figure 7.1). In terms of data management, it is important to assign and document roles, not just presume roles, for many reasons, including the following (UK Data Service 2023):

1. It allows team members to begin standardizing workflows.
2. When team members know exactly what is expected of them, it keeps data more secure.
3. Creating contingency plans for when staff can no longer fulfill their roles allows for the continuity of practices.

7.1 Research Project Roles

Before diving in to how to assign and document roles for a project, it is important to get an understanding of typical roles on an education research project team (Elsevier Author Services 2021). Your team may be lucky enough to have all, or several of these roles. Other times, just one person, such as the principal investigator, may take on all or multiple of these roles. With that said, if your budget allows it, I highly recommend hiring individuals to fill each of the roles mentioned below to allow team members to specialize and excel in their area of expertise. While learning all aspects of a project is highly recommended to create a cohesive team that works collaboratively, team members who take on too many project roles can be spread too thin and project goals may suffer.

7.1.1 Investigators

The investigators, also known as PIs (principal investigators) and co-PIs, are the individuals who prepare and submit the grant proposal and are responsible for the administration of that grant. There is often more than one investigator on a project including someone with content area knowledge, as well as a methodologist. PIs and Co-PIs have varying levels of involvement in

DOI: 10.1201/9781032622835-7

research projects and are typically, not always, more hands off in the day-to-day administration. Even if some tasks are delegated to other research staff, PIs and Co-PIs are ultimately responsible for meeting grant requirements, as well as requirements from other oversight departments or partners (e.g., Institutional Review Board (IRB) submissions, agreement compliance, effort reporting, progress reporting) (Washington University in St. Louis 2023).

7.1.2 Project Coordinator

The project coordinator (or project manager) is an essential member of the research team. As the name implies, this person typically coordinates all research activities and ensures compliance with oversight such as the IRB. Tasks they may oversee include recruitment and consenting of participants, creation of data collection materials, creation of protocols and standard operating procedures (SOPs), training data collectors, data collection scheduling, and more. The project coordinator may also supervise many of the other research team roles, such as research assistants.

7.1.3 Data Manager

The data manager is also an essential member of the team. This person is responsible for the organizing, cleaning, documenting, storing, and dissemination of research project data. This team member works closely with the project coordinator, as well as the investigators, to ensure that data management is considered throughout the project life cycle. Tasks a data manager may oversee include data storage, security and access, building data collection and tracking tools, data cleaning and validation, data documentation, and organizing data for sharing purposes.

This role is vital in maintaining the standardization of data practices. If you do not have the budget to hire a data manager, make sure to assign someone on your team to oversee the flow of data, ensuring that throughout a project, data is documented, collected, entered, cleaned, and stored consistently and securely.

7.1.4 Project Team Members

This role refers to any staff hired to help implement a research project which may include full-time staff members, with titles such a research or project assistants for instance, or it may include part-time team members, such as graduate students. Project team members are often out in the field, collecting data, or they may also assist in other areas such as preparing data collection materials, entering data, or assisting with data management. Senior project team members may also assist in implementing training or acting as data collection leads in the field.

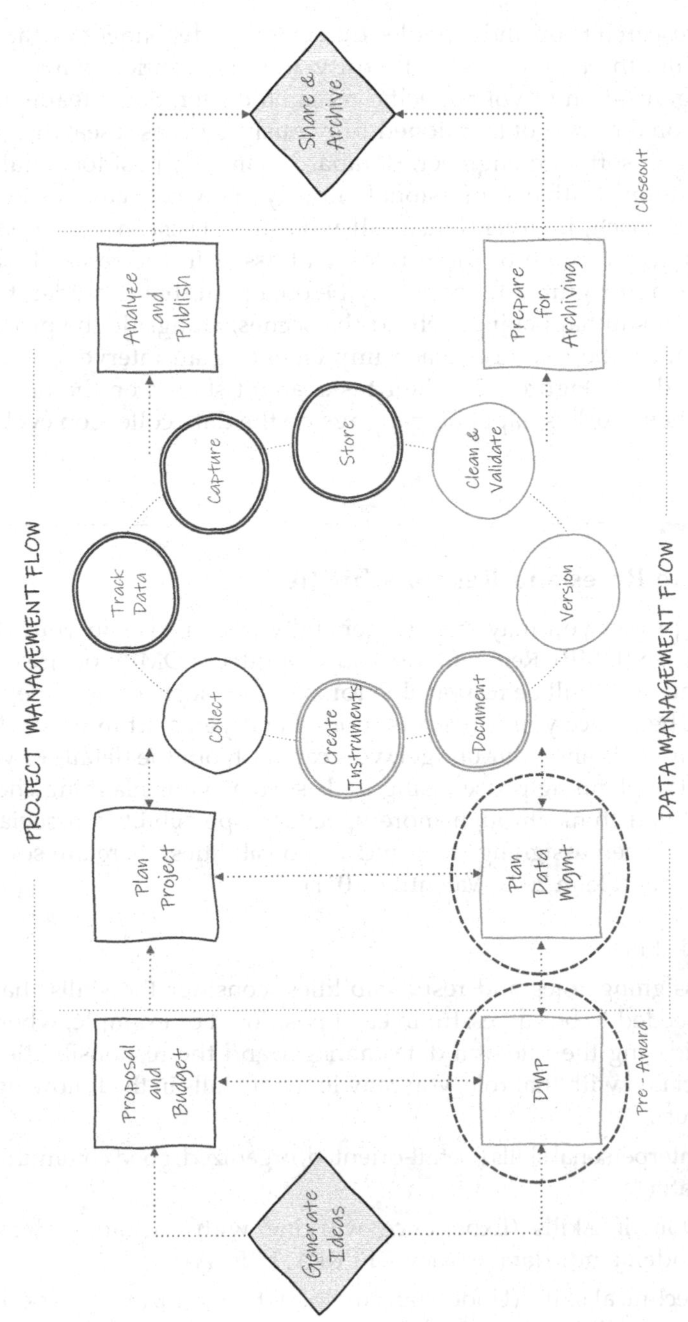

FIGURE 7.1
Planning roles in the research project life cycle.

7.1.5 Other Roles

The size of a research team and the roles that exist are dependent on factors such as funding, the type of research study, the intervention being studied, or the organization of your specific research center. Some teams may include additional roles, not mentioned previously, such as research director, lab manager, software engineer, database manager, postdoc, analyst, statistician, administrative professional, hourly data collector, outreach coordinator, or coach/interventionist, all who may assist in the research cycle in other ways. Some of these roles will assist in the research data life cycle as seen in Figure 7.1. Some may be on a path that is hidden from the diagram but still happening, behind the scenes, alongside the process. Take, for instance, the role of a coach implementing an intervention that is being studied (see Figure 7.2). Their tasks aren't shown on the original diagram but their work is happening alongside the data collection cycle.

7.2 Assigning Roles and Responsibilities

Early on in a project you may start to generally assign roles in your data management plan (DMP). Remember if you submitted a DMP, you are often required to state who will be responsible for activities such as data integrity and security. Then, once your project is funded and you start to have a better idea of your goals and your budget, you can flesh out the details of your roles. During the planning phase, using tools such as your planning checklists will help you think through more specific responsibilities associated with each role. When assigning roles and responsibilities, there are several factors to consider (Helm 2022; Valentine 2011).

1. Required skill set
    - In assigning roles and responsibilities, consider the skills that are needed to be successful in each position. For example, when considering the role of a data manager and the responsibilities associated with that role, you may look for skills in the following buckets:
        - Interpersonal skills (Detail-oriented, organized, good communicator)
        - Domain skills (Experience working with education data, understands data privacy—FERPA, HIPAA)
        - Technical skills (Understanding of data organization, experience building data pipelines, coding experience, specific software/tool experience)

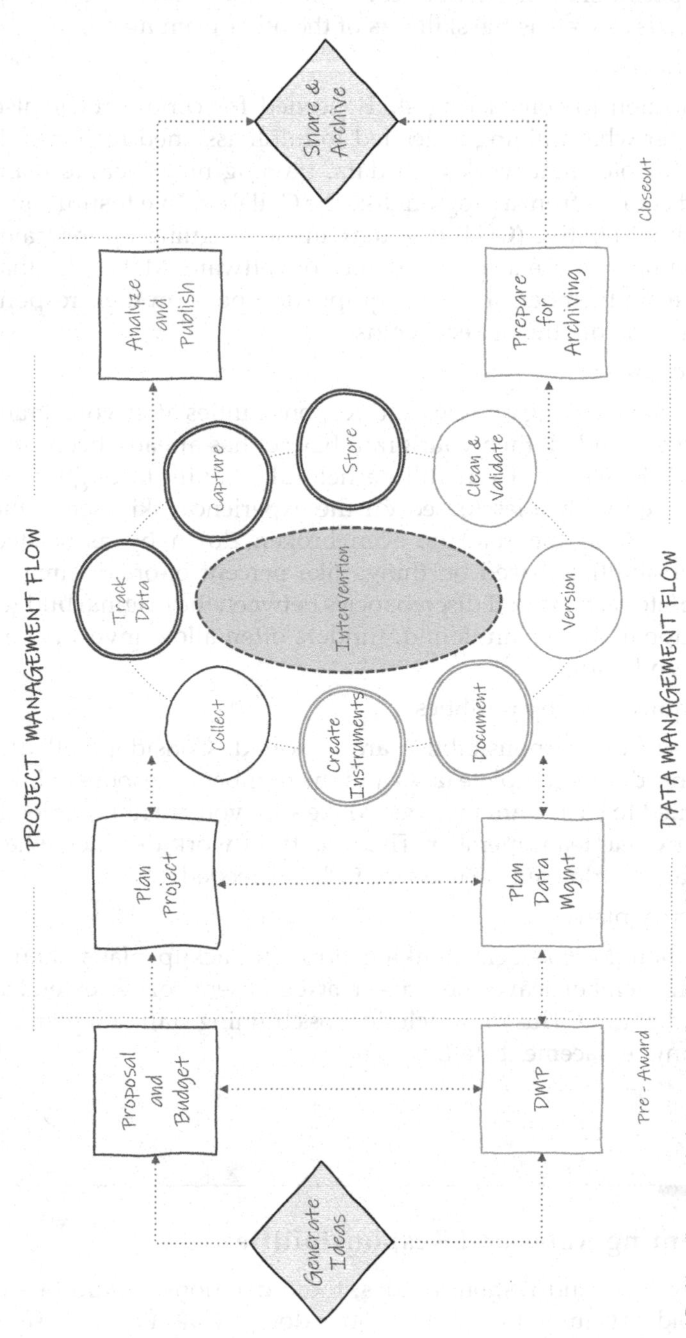

* Double outline means both teams typically collaborate on this phase

FIGURE 7.2
Life cycle diagram updated to show hidden processes.

- The specific skills needed for each role will depend on your project needs as well as the skill sets of the other team members.

2. Training needs
   - In addition to considering skills needed for certain roles, also consider what training is needed to fulfill assigned responsibilities. In roles that work with data, training may include mandated courses from a program like the Collaborative Institutional Training Initiative (CITI) or it may involve signing up for training on how to use a specific device or software. Make sure that your team members are well-equipped to perform their responsibilities before the project begins.
3. Estimated costs
   - If you are working on roles and responsibilities after your grant has been funded, then your grant budget has already been submitted. However, it can still be helpful to think through costs associated with roles (based on the experience/skill set of the person filling the role) or even broken down by associated responsibilities (based on things like percent effort or time to complete each task). If discrepancies between the original budget and updated costs are found, funders often allow investigators to amend budgets.
4. Assess equity in responsibilities
   - Review how responsibilities are allocated. Consider both the time needed to complete tasks and the number of responsibilities assigned to each team member. Make sure you are not overloading any one team member. The quality of work declines when staff are overloaded, so reassign tasks as needed.
5. Contingency plans
   - You should also begin thinking through backup plans should a staff member leave the project or be absent for an extended period of time. This may include cross training staff or a plan for training replacement staff.

7.3 Documenting Roles and Responsibilities

After assigning roles and responsibilities, those decisions should be documented to avoid any ambiguity about who is doing what. While these roles may be briefly documented in a data management plan, it is important to more thoroughly document this information as well.

There are many reasons to document staff roles and responsibilities and to store that information in a central, accessible location.

1. It allows your team to easily reference the document to see who is on the project team, what roles they play, and who to contact for questions regarding various project aspects (e.g., who to contact for data storage access).
2. As new tasks arise, team members can refer to the document to see who is best fit for the assignment.
3. Last, reviewing roles and responsibilities in a document also helps you more clearly see what responsibilities are assigned and how they are assigned. After reviewing the document, you can make further revisions if responsibilities need to be added or further redistributed in any way.

Figure 7.3 is one example of a roles and responsibilities document that organizes information by roles. Note that this document only lists overarching

Project Name:
Date:

The purpose of this document is to clearly articulate the different roles within a project team and the duties each role/person is responsible for.

Title	Role	Name
Project Coordinator	Oversee the completion of research study objectives and research compliance	(Name of Individual)
Responsibilities		
• Hire and train part-time data collectors • Recruit and consent study schools and teachers • Consent study students • Build data collection tools • Organize data collection efforts • Document data collection efforts		

Title	Role	Name
Data Manager	Oversee the design of data collection tools, data documentation as well as the security, management and integrity of study data	(Name of Individual)
Responsibilities		
• Monitor data training compliance for all staff • Build data collection tools • Build data tracking database • Document data management efforts • Clean study data • Oversee data access • Oversee data storage		

FIGURE 7.3
Roles and responsibilities document organized by role.

Project Name:
Date:

The purpose of this document is to clearly articulate the different roles within a project team and the duties each role/person is responsible for.

Phase	Project Coordinator [Name]	Data Manager [Name]	Research Assistant [Name]
Documentation	• Create documentation	• Create documentation	
Create Instruments	• Build data collection instruments • Order supplies	• Build participant tracking database	• Test data collection tools and provide feedback
Data Collection	• Hire data collectors • Train data collectors • Schedule data collection • Monitor quality control • Track collection	• Oversee integrity of data	• Collect data
Data Capture	• Oversee entry of data	• Build data entry database	• Enter data
Data Storage	• Ensure raw electronic data is stored correctly • Ensure paper data is stored securely in the office	• Manage data access • Oversee data storage backup and security	

FIGURE 7.4
Roles and responsibilities organized by phase.

responsibilities, not detailed steps associated with tasks. Specific actionable steps will be laid out in other process documentation such as standard operating procedures (see Section 8.2.7), where names are attached to each task.

However, this document can be laid out in any format that conveys the information clearly to your team. Figure 7.4 is an example of organizing information by phases of the research life cycle.

Last, any additional information can be added to these documents to help clarify responsibilities, such as:

- Links to related standard operating procedures (e.g., for building a participant tracking database you may link to the specific SOP that lays out steps for building the tool).

- Names of other staff members (if any) who assist with or also contribute to each responsibility.
- Timing of each responsibility (e.g., weekly, ongoing, the month of February).
- Name of team members who will take on responsibilities in case of a team member's absence.

8

Documentation

Documentation is a collection of files containing procedural and descriptive information about your team, project, workflows, and data. Creating thorough documentation during your study is equally as important as collecting your data. Documentation serves many purposes including:

- Standardizing procedures
- Strategic planning
- Securing data and protecting confidentiality
- Tracking data provenance
- Discovering errors
- Enabling reproducibility
- Ensuring others use and interpret data accurately
- Providing searchability through metadata

This chapter will cover four levels of documentation: Team, project, dataset, and variable levels. Even though we are currently discussing the documentation phase, some of these documents will be created earlier or later than this phase, and the timing will be discussed in each section. During a project, while you are actively using your documents, the format of these documents does not matter. Choose a human-readable format that works well for your team (Word, PDF, TXT, Google Doc, XLSX, HTML, OneNote, etc.). With that said, you will want to have a long-term plan for converting these documents, as needed, into formats that are both interoperable and sustainable for archiving and sharing (see Chapter 15 for more information).

The documents below are all recommended and will help you successfully run your project. You can create as many or as few of these documents as you wish. The documents you choose to produce should be based on what is best for your project and your team, as well as what is required by your funder and other governing bodies. No matter which documents you choose to implement, it is important to create templates for your documents and implement them consistently within, and even across projects. Creating documentation using templates, or consistent formats and fields, reduces

 DOI: 10.1201/9781032622835-8

duplication in efforts (no need to reinvent the wheel) and allows your team to interpret documents more easily. These documents are best created by the team member who directly oversees the process, and sometimes that may include a collaborative effort (for example, both a project coordinator and a data manager may build documents together). Ultimately, though, all your documents should be reviewed with your data management working group (DMWG) in order to gather feedback and reach consensus (see Chapter 6).

Each type of documentation discussed below is a living document to be updated as procedures change, or new information is received. As seen in the cyclical section of Figure 8.1, team members should revisit documentation each time new data is collected, or more often if needed, to ensure documentation still aligns with actual practices. If changes are made and not added to documentation over long periods of time, you will find that you no longer remember what happened and that information will be lost. It will also be important to version your documents (i.e., track major changes) along the way so that staff know that they are working with the most recent version and can see when and why documents have been updated.

NOTE

Creating and maintaining these documents **is an investment**. However, the return on investment is well worth the effort. Make sure to account for this time and expertise in your proposal budget (see Section 5.5).

8.1 Team-Level

Team-level data management documentation typically contains data governance rules that apply to the entire team, across all projects. While these documents can be amended any time, they should be started long before you apply for a grant, when your lab, center, or institution is formed (see Figure 8.2).

8.1.1 Lab Manual

When to create: Team document phase

A lab manual, or team handbook, creates common knowledge across your team (Mehr 2020). It provides staff with consistent information about lab culture—how the team works and why they do the things they do. It also sets

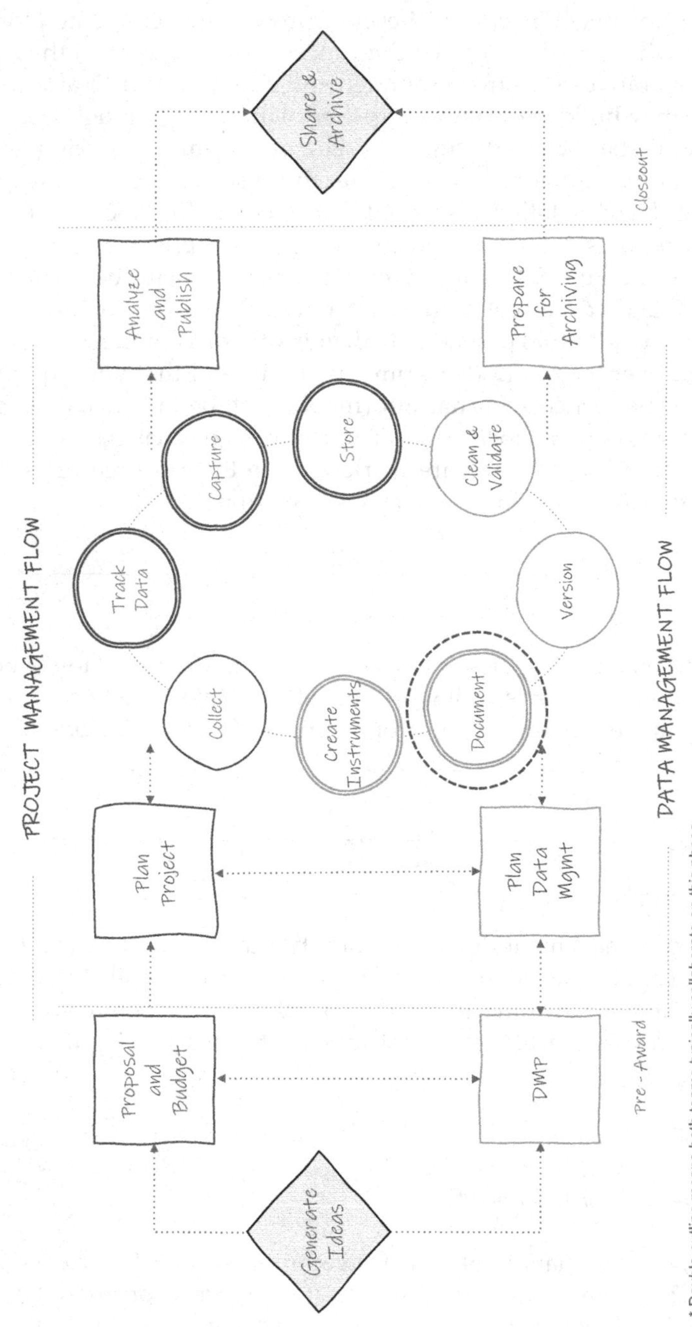

FIGURE 8.1
Documentation in the research project life cycle.

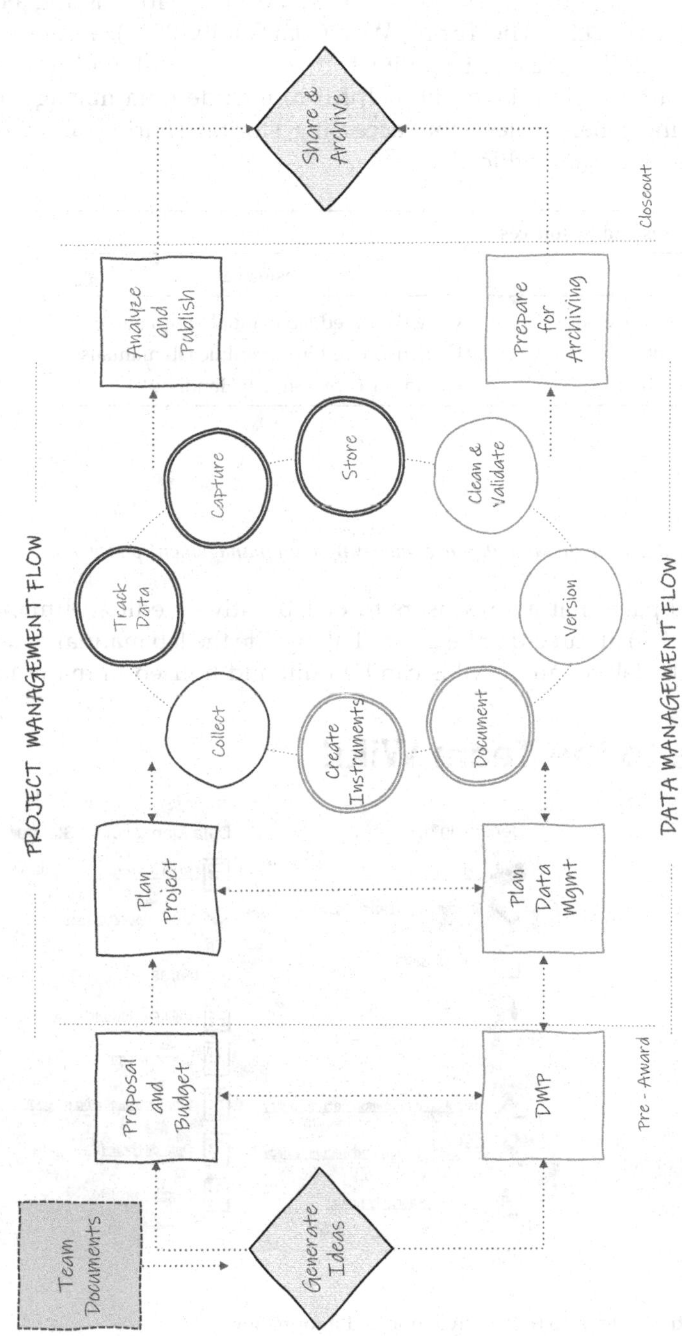

FIGURE 8.2
Team-level documentation in the research project life cycle.

expectations, provides guidelines, and can even be a place for passing along career advice (Aczel 2023; The Turing Way Community 2022). While a lab manual will primarily consist of administrative, procedural, and interpersonal types of information, it can be helpful to include data management content, including general rules about accessing, storing, sharing, and working with data securely and ethically.

Templates and Resources	
Source	**Resource**
Balazs Aczel, et al.	Crowdsourced lab manual template[1]
Hao Ye, et al.	Crowdsourced list of public lab manuals[2]
Samuel Mehr	Common Topics in Lab Handbooks[3]

8.1.2 Wiki

When to create: Team document phase, planning data management phase

A wiki is a webpage that allows users to collaboratively edit and manage content (Figure 8.3). It can either be created alongside the lab manual or as an alternative to the lab manual. Wikis can be built and housed in many tools

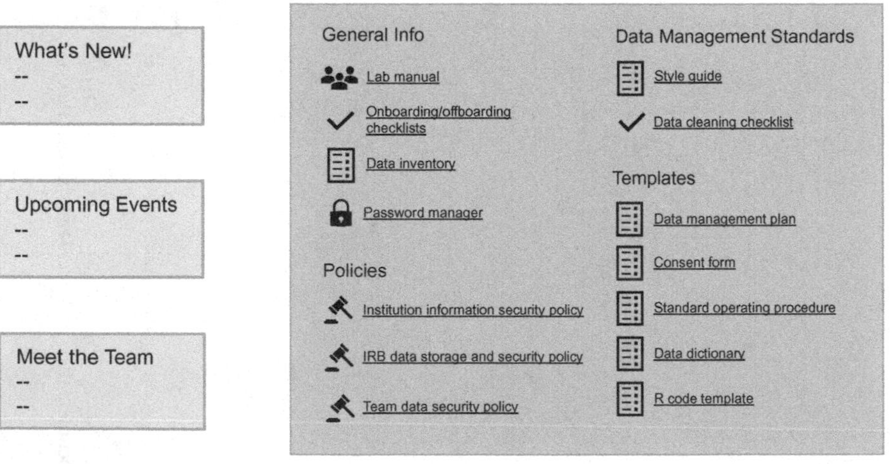

FIGURE 8.3
Example team wiki with links to frequently requested information.

(e.g., SharePoint, Teams, Notion, GitHub). While some lab wikis are public, most are not and can be restricted to invited users only. Wikis are a great way to keep disparate documents and pieces of information, for both administrative and data related purposes, organized in a central, accessible location. Your wiki can include links to important documents and policies, or you can also add text directly to the wiki to describe certain procedures. Rather than sending team members to multiple different folders for frequently requested information, you can refer them to your one wiki page.

NOTE

Project-level wikis can also be created and are useful in centralizing frequently referenced information pertaining to specific projects.

Templates and Resources

Source	Resource
Aly Lab	Example public lab wiki[4]
Notion	How to build a wiki for your company[5]
SYNC Lab	Example public lab wiki[6]

8.1.3 Onboarding and Offboarding

When to create: Team document phase

While an **onboarding** checklist will mainly consist of non-data-related, administrative information such as how to sign up for an email or how to get set up on your laptop, it should also contain several data-specific pieces of information to get staff generally acclimated to working with data in their new role. This ensures that everyone is receiving consistent information about data practices and standards (Briney et al. 2020).

Similarly, while an **offboarding** checklist will contain a lot of procedural information about returning equipment and handing off tasks, it should also contain information specific to data management and documentation that help maintain data integrity and security.

Data-related topics to consider adding to your onboarding and offboarding checklists are included in Figure 8.4.

Onboarding	Offboarding
☐ Contacts ☐ For acquiring access to data storage spaces ☐ For data related questions ☐ Learning ☐ Where to go to learn more about existing data for current and past projects (e.g., data inventory) ☐ Relevant standard operating procedures to review ☐ Where to review roles and responsibilities ☐ Requirements and standards ☐ Any required training (e.g., CITI, IT training) ☐ Any required documents to review and complete (e.g., data security policy) ☐ Any standards to review (e.g., style guide) ☐ Tools ☐ What existing data tools are used ☐ How to access those tools ☐ Training needed for those tools	☐ Access ☐ Contacts for removing data access ☐ Tying up loose ends ☐ Making sure all standard operating procedures associated with your role are up to date ☐ Review all data files you have worked on to ensure they are ☐ Stored according to policy ☐ Documented adequately ☐ Named according to the style guide ☐ Accessible to someone on the team other than yourself

FIGURE 8.4
Sample data topics to add to onboarding and offboarding checklists.

Template and Resources

Source	Resource
Crystal Lewis	Sample data topics to add to an onboarding checklist[7]
Crystal Lewis	Sample data topics to add to an offboarding checklist[8]

8.1.4 Data Inventory

When to create: Team document phase, prepare for archiving phase

A data inventory maps all datasets collected by the research team across all projects (Salfen 2018; Van den Eynden et al. 2011) (Figure 8.5). As a team

Project	Dataset Name	File size	Description	Linking variables
Project A: Description of project (2018-2021)	pa_stu_c1-c3_clean.sav	N = 680 # of vars = 140	Student survey, assessment, and school records data Cohorts 1-3	stu_id, tch_id
	pa_tch_c1-c3_clean.sav	N = 110 # of vars = 92	Teacher survey and classroom observation data Cohorts 1-3	tch_id
Project B: Description of project (2016-2018)	pb_tch_c1-c2_clean.sav	N = 63 # of vars = 41	Teacher survey data Cohorts 1-2	tch_id, sch_id
	pb_sch_c1-c2_clean.sav	N = 8 # of vars = 25	School enrollment and demographic data Cohorts 1-2	sch_id

FIGURE 8.5
Example data inventory document.

grows, the number of datasets typically expands as well. It can be very helpful to keep an inventory of what datasets are available for team members to use, as well as details about those datasets. Types of information to share in a data inventory include:

- Project associated with each dataset
- Dates that each dataset was collected
- Details about each dataset (what the dataset contains, how it is organized, what questions can be answered with the data)
- How datasets are related
- Links to relevant documentation
- Storage location of each dataset or who to contact for access

8.1.5 Team Data Security Policy

When to create: Team document phase

A team-level data security policy is a set of formal guidelines for working with data within an organization. While this policy may draw from information in broader institutional data security policies, this team-level policy is written specifically for your group and should broadly cover how team members are allowed to work with data in a way that protects research participant privacy, ensures quality control, and adheres to legal, ethical, and technical guidelines. Documenting this information ensures a cohesive understanding among team members regarding the terms and conditions of project data use (CESSDA Training Team 2017). A data security policy can be added to a lab manual, or created as a separate document where team members can even sign (Filip 2023) or check a box acknowledging that they have read and understand the policy.

Ideas of content to include in a team-level data security policy are included in Figure 8.6.

Templates and Resources

Source	Resource
Crystal Lewis	Data security policy template[9]
SYNC Lab	Data security protocol[10]
University of Nebraska-Lincoln	Research data and security checklist[11]

• Requirements o What is required before staff can work with data? (i.e., CITI training, IT security training, signing this agreement) o Review relevant information that impacts how data is managed (e.g., data sensitivity levels, FERPA, IRB policies, DUAs, quality control concerns) • Data storage and access o Electronic data ▪ Where is it stored? ▪ How is it secured? ▪ Who has access? ▪ How are the files organized? ▪ How is data backed up? ▪ How long is it retained? When is it destroyed? o Paper data ▪ Where is it stored? ▪ How is it secured? ▪ Who has access? ▪ How are files organized? ▪ How long is it retained? When is it destroyed?	• Working securely with data o Electronic data ▪ What are the rules for working with electronic data securely? ▪ What are the rules for securing devices? ▪ What are the rules for transmitting electronic data securely? o Paper data ▪ What are the rules for working with paper data securely? • In the field • In the office • Analyses o What are the rules for using research project data for personal or project analyses? (e.g., request process required) • Contacts o Who are the contacts for all data access needs? o Who are the contacts for questions or concerns? (e.g., confidentiality breaches, errors found in the data)

FIGURE 8.6
Example of content to include in a team data security policy.

8.1.6 Style Guide

When to create: Team document phase, planning data management phase

A style guide is a set of standards for the formatting of information. It improves consistency and creates a shared understanding within and across files and projects. This document includes conventions for procedures such as variable naming, variable value coding, file naming, file structure, and even coding practices. It can be created in one large document or separate files for each type of procedure. I highly recommend applying your style guide consistently across all projects which is why this is included in the team documentation. Since style guides are so important, and there are many recommended practices to cover, I have given this document its own chapter. See Chapter 9 for more information.

Templates and Resources

Source	Resource
Hadley Wickham	Example R coding style guide[12]
Strategic Data Project	Example style guide[13]

8.2 Project-Level

Project-level documentation is where all descriptive information about your project is contained, as well as any planning decisions and process documentation specifically related to your project.

8.2.1 Data Management Plan

When to create: DMP phase

As discussed in Chapter 5, if your project is federally funded it is likely that a data management plan is required. This project-level summary document is created in the DMP phase, long before a project begins. However, your DMP can continue to be modified throughout your entire study. If any major changes are made, it may be helpful to reach out to your program officer to keep them in the loop as well.

8.2.2 Data Sources Catalog

When to create: DMP phase

Also, as reviewed in Section 5.3, a data sources catalog is an excellent project planning tool that should be developed early on during the DMP phase. This spreadsheet helps you succinctly summarize the data sources you will collect for your project, as well as plan the details of how data will be collected, managed, and shared. This document serves as a referral source for the remaining planning phases of your project and should be updated as needed.

8.2.3 Checklists and Meeting Notes

When to create: DMP phase, planning data management phase

Checklists, as discussed in Section 6.3, are documents that are created, or copied from existing templates, and reviewed during the planning phase. Using checklists facilitates discussion and allows your team to build a cohesive understanding for how data will be managed throughout your entire project. As you work through the checklists, all decisions made should be documented in meeting notes. Consider creating and using a template to standardize the flow of your meeting notes across the various team members who may take notes.

Example meeting notes template.

```
Date: YYYY-MM-DD
Attendees: Name1, Name2, Name3

Agenda Items:
  1. Topic 1
  2. Topic 2
  3. Topic 3

Action Items:
  Name1:
    - Tasks
  Name2:
    - Tasks
  Name3:
    - Tasks
```

All meeting notes should be stored in a central location where team members can reference them as needed (e.g., a planning folder with notes ordered by date, a running document linked on a project wiki page). After the planning phase is complete, any decisions should be formally documented in applicable team-, project-, data-, or variable-level documents (e.g., research protocol, SOPs, style guide, or roles and responsibilities documents). Even beyond the planning phase, though, all meeting decisions and discussions should continue to be documented in meeting notes and used to update formal documentation as needed.

8.2.4 Roles and Responsibilities Document

When to create: DMP phase, planning data management phase

Using the checklists reviewed during the DMP and planning phase, your team should begin assigning roles and responsibilities for your project. Those designations should be formally documented and shared with the team. In Chapter 7 we reviewed ways to structure this document. Once this document is created, make sure to store it in a central location for easy referral and update the document as needed.

Templates and Resources

Source	Resource
Crystal Lewis	Roles and responsibilities template organized by phase in wide format[14]
Crystal Lewis	Roles and responsibilities template organized by role[15]

8.2.5 Research Protocol

When to create: Documentation phase

The research protocol is a comprehensive project plan that describes the what, who, when, where, and how of your study. Many of the decisions made while writing your data management plan and reviewing your planning checklists will be summarized in this document. If you are submitting your study to an Institutional Review Board, you will most likely be required to submit this document as part of your application. A research protocol assists the board in determining if your methods provide adequate protection for human subjects. In addition to serving this required purpose, the research protocol is also an excellent document to share along with your data at the time of data sharing, and an excellent resource for you when writing technical reports or manuscripts. This document provides the context needed for you and others to effectively interpret and use your data. Make sure to follow your institution's specific template if provided, but items commonly included in a protocol are shown in Figure 8.7.

When it comes time to deposit your data in a repository, the protocol can be revised to contain information helpful for a data end user, such as known limitations in the data. Content such as risks and benefits to participants might be removed, and numbers such as study sample count should be updated to show your final numbers. Additional supplemental information can also be added as needed (see Section 8.2.6).

✓ Funding source ✓ Overview of study ✓ Intervention and research design ✓ Setting and sample (including anticipated numbers) ✓ Anticipated benefits and risks to participants ✓ Participant compensation ✓ Project timeline (what data will be collected, on whom, and when)	✓ Measures used in study (including citations and versions) ✓ Overview of study procedures (recruitment, consent, inclusion/exclusion criteria, randomization, data collection) ✓ Data preparation and processing plan (data safety monitoring, data storage, data quality monitoring, de-identification, data sharing) ✓ Handling unexpected events ✓ Data analysis plan

FIGURE 8.7
Common research protocol elements.

Templates and Resources

Source	Resource
Crystal Lewis	A template to create a project-level summary document for data sharing (based on an IRB research protocol)[16]
Jeffrey Shero, Sara Hart	IRB protocol template with a focus on data sharing[17]
The Ohio State University	Protocol template[18]
University of Missouri	Protocol template[19]
University of Washington	Protocol checklist[20]

8.2.6 Supplemental Documents

While these documents can absolutely stand alone, I am calling these supplemental documents because they can also be added to your research protocol, at any point, as an addendum to further clarify specifics of your project.

8.2.6.1 Timeline

When to create: Planning data management phase

The first supplemental document that I highly recommend creating is a visual representation of your data collection timeline. When you first create a timeline, it will be based on your best estimate of the time it will take to complete milestones, but like all documents we've discussed, it can be updated as you learn more about the reality of the workload. This document can be both a helpful planning tool (for both project and data teams) in preparing for times of heavier and lighter workloads, as well as an excellent document to share with future data users to better understand waves of data collection. There is no one format for how to create this document. Figure 8.8 is an example of one way to visualize a data collection timeline.

8.2.6.2 Participant Flow Diagram

When to create: After data collection

A participant flow diagram displays the movement of participants through a study, assisting researchers in better understanding milestones such as eligibility, enrollment, and final sample counts. As seen in Figure 8.9, these diagrams are helpful for assessing study attrition and reasons for missing data can be described in the diagram. In randomized controlled trial studies, these visualizations are more formally referred to as CONSORT (Consolidated

	Year 1									Year 2								
	Sep	Oct	Nov	Dec	Jan	Feb	Mar	Apr	May	Sep	Oct	Nov	Dec	Jan	Feb	Mar	Apr	May
Cohort 1 student survey																		
Cohort 1 teacher survey																		
Cohort 1 student assessment																		
Cohort 2 student survey																		
Cohort 2 teacher survey																		
Cohort 2 student assessment																		

FIGURE 8.8
Example data collection timeline.

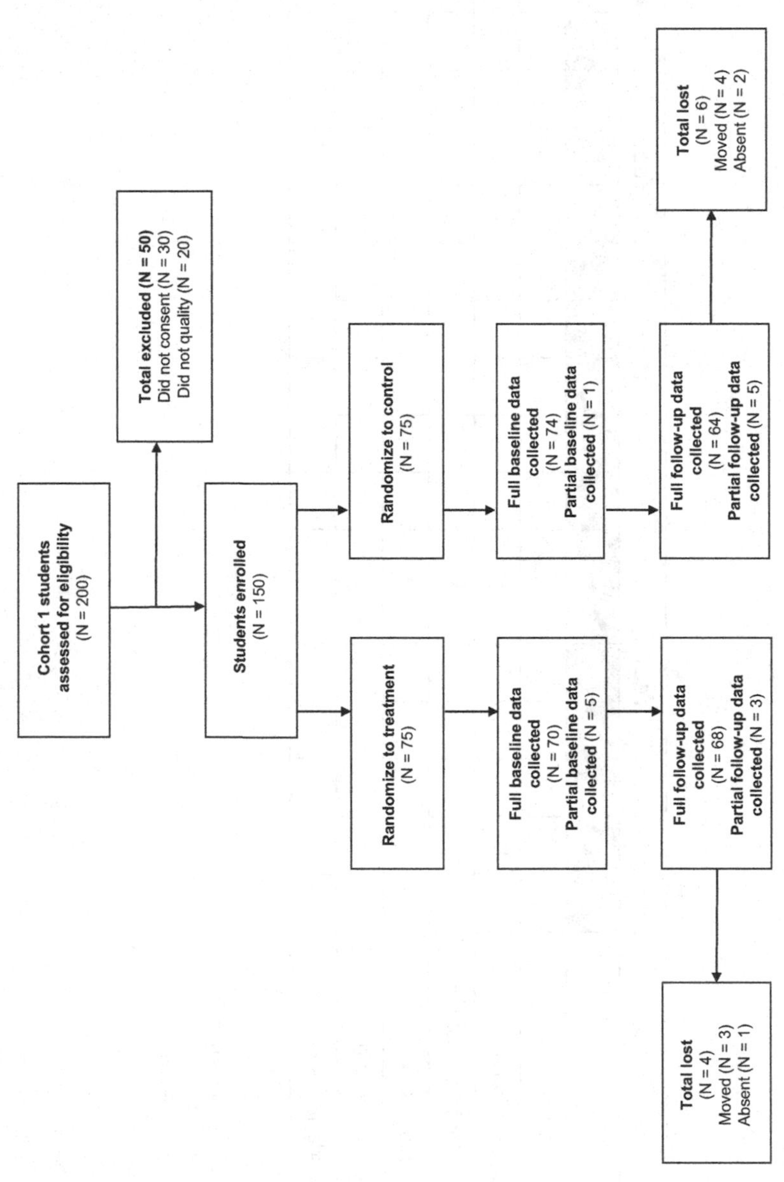

FIGURE 8.9
Example participant flow diagram.

Standards of Reporting Trials) diagrams (Schulz et al. 2010). They provide a means to understand how participants are randomized and assigned to treatment groups. As you can imagine, though, this diagram cannot be started until participants are recruited and enrolled, and it must be updated as each wave of data is collected. Your participant tracking database, which we will discuss in Chapter 10, will inform the creation of this diagram.

8.2.6.3 Instruments

When to create: Create instruments phase

Unless a form is proprietary, actual copies of instruments can be included as supplemental documentation. This includes copies of surveys, assessments, and other forms. It can also include any technical documents associated with your instruments or measures (i.e., a technical document for an assessment or a publication associated with a measure you used). Sometimes researchers will annotate instruments to show how items were named or coded (see Figure 11.5).

8.2.6.4 Flowchart of Data Collection Instruments

When to create: Create instruments phase

You can also include flowcharts of how participants are provided or assigned to different instruments or screeners to help users better understand issues such as missing data (see Figure 8.10).

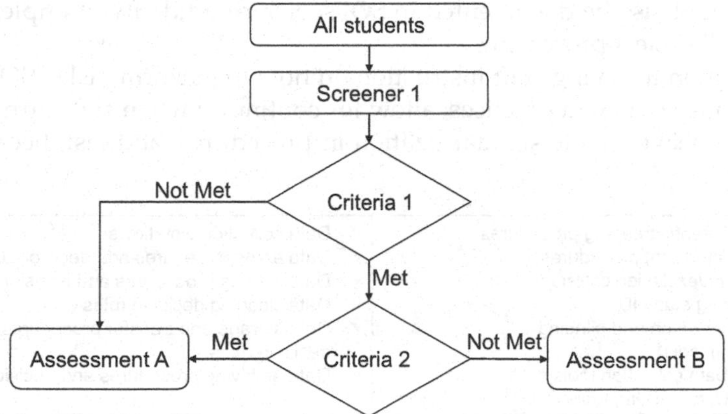

FIGURE 8.10
Example instrument decision making flowchart.

8.2.6.5 Consent Forms

When to create: Create instruments phase

Informed consent forms (see Section 11.2.5) can also be added as an addendum to research protocols to give further insight into what information was provided to study participants.

8.2.7 Standard Operating Procedures

When to create: Documentation phase

While the research protocol provides summary information for all decisions and procedures associated with a project, we still need documents to inform how the procedures are actually implemented on a daily basis (NUCATS 2023). Standard operating procedures (SOPs) provide a set of detailed instructions for routine tasks and decision-making processes. If you recall from Chapter 6, every step we added to a data collection workflow is then added to an SOP, and the details fleshed out. Not only will you have an SOP for each source of data you are collecting (i.e., teacher survey, student assessments, teacher observation), you should also have SOPs for any other decisions or processes that need to be repeated consistently or followed in a specific way to maintain compliance (Hollmann et al. 2020). Many of the decisions laid out in your protocol will be further detailed in an SOP. Examples of data management procedures to include in an SOP are provided in Figure 8.11. Additional project management tasks such as recruitment procedures, personnel training, data collection scheduling, or in-field data collection routines, should also be documented in SOPs, ensuring fidelity of implementation for all project procedures.

In addition to giving staff instruction on how to perform tasks, SOPs also create transparency in practices, allow for continuity when staff turnover or go out on leave, create standardization in procedures, and last, because an

<table>
<tr>
<td>✓ Recruitment/screening procedures
✓ Consent/assent procedures
✓ Inclusion/exclusion criteria
✓ Assigning study IDs
✓ Randomization and blinding
✓ Building tools
    ✓ Data collection tools
    ✓ Data tracking tools</td>
<td>✓ Data collection workflows
✓ Data entry procedures and decision rules
✓ Data scoring procedures and decision rules
✓ Data cleaning decision rules
✓ Data storage and transfer procedures and decision rules
✓ Data archiving procedures and decision rules</td>
</tr>
</table>

FIGURE 8.11

Examples of data management SOPs to create.

SOP should include versioning information, they allow you to accurately report changes in procedures throughout the project. You will want to create a template that is used consistently by all staff who write SOPs (see Figure 8.12 for an example).

In developing your SOP template, you should begin with **general information** about the scope and purpose of the procedure, as well as any relevant tools, terminology, or documentation. This provides context for the user and gives them the background to use and interpret the SOP. The next section in the SOP template, **procedures**, lists all steps in order. Each step provides the name of the staff member/s associated with that activity to ensure no ambiguity. Steps should be as detailed as possible so that you could hand your SOP over to any new staff member, with no background in this process, and be confident they can implement the procedure with little trouble. Specifics such as names of files and links to their locations, names of contacts, or methods of communication (e.g., email vs instant message) should be included. Additions such as screenshots, links to other SOPs or workflow diagrams, or even links to online tutorials or staff-created how-to videos can also be

Title			
Who Created			
Creation Date		Version Number	

General Information

1. Purpose: What functions does this SOP cover?
2. Scope: What project/s does this SOP apply to?
3. Technology required: What tools are required to implement this SOP?
4. Terminology and abbreviations used: Define any unclear terms or acronyms used in this SOP.
5. Related documentation: Link to any related documents, videos, or tutorials that may help users interpret this SOP.
6. Applicable policies: Link to any applicable regulations, guidelines, or policies.

Procedures (in order):

1. [Name of person responsible]: Name of step
   a. Detailed associated steps
   b. Screenshots as needed
   c. Links to associated documents, videos, tutorials
2. [Name of person responsible]: Name of step
   a. Detailed associated steps
   b. Screenshots as needed
   c. Links to associated documents, videos, tutorials

Revision History

Version Number	Revision Date	Description of and Reason for Revision	Who Created Revision

FIGURE 8.12
Standard operating procedure minimal template.

embedded. Last, any time revisions are made to the SOP, clarifying information about the update is added to the **revision** section and a new version of the SOP is saved (see Section 9.3 for ideas on how to version a file name). This allows you to keep track of what changes were made over time, including when they were made and who made them.

Templates and Resources	
Source	**Resource**
Crystal Lewis	SOP template[21]

8.3 Dataset-Level

Dataset-level documentation applies to your datasets and includes information about what they contain and how they are related. It also captures things such as planned transformations, potential issues to be aware of, and any alterations made to the data. In addition to being helpful descriptive information, a huge reason for creating dataset documentation is authenticity. Datasets go through many iterations of processing which can result in multiple versions of a dataset (CESSDA Training Team 2017; UK Data Service 2023). Preserving data lineage by tracking transformations and errors found is key to ensuring that you know where your data come from, what processing has already been completed, and that you are using the correct version of the data.

8.3.1 Readme

When to create: At any time they are useful

A README is a plain text document that contains information about your files. These stem from the field of computer science but are now prevalent in the research world. These documents are a way to convey pertinent information to collaborators in a simple, no-frills manner. READMEs can be used in many different ways, but I will cover three ways they are often used in data management.

1. For conveying information to your colleagues
    - Imagine a scenario where a study participant reaches out to a project coordinator to let them know that they entered the

incorrect ID in their survey. When the project coordinator downloads the raw data file to be cleaned by the data manager, they also create a file named "readme.txt" that contains this information and it is saved alongside the data file in the raw data folder. Now, when the data manager goes to retrieve the data file, they will see that a README is included and know to review that document first.

Example README used to convey information about a file

```
- ID 5051 entered incorrectly. Should be 5015.
- ID 5089 completed the survey twice
  - First survey is only partially completed
```

2. For conveying steps in a process (sometimes also called a setup file)
   - There may be times that a specific data pipeline or reporting process requires multiple steps, opening different files and running different scripts. This information could go in an SOP, but if it is a programmatic-type process, completed using a series of scripts, it might be easiest to put a simple file named "readme_setup.txt" in the same folder as your scripts so that someone can easily open the file to see what they need to run.

Example README used to convey information about steps in a process

```
  Step 1: Run the file 01_clean_data.R to clean the data
  Step 2: Run the file 02_check_errors.R to check for
errors
  Step 3: Run the file 03_run_report.R to create report
```

3. For providing information about a set of files in a directory
   - It can be helpful to add a README to the top of your directories when both sharing data internally with colleagues, or when sharing files in an external repository. Doing so can provide information about what datasets are available in the directory and pertinent information about those datasets. This README can include things like a description of the files, how the datasets are related and can be linked, information associated with different versions, definitions of common prefixes, suffixes, or acronyms used in datasets, or instrument response rates. Figure 8.13 is an example README that can be used to describe all data sources shared in a project repository.

File Name	Description	Record Level	N	# of Vars	Linking IDs	Prefixes/Suffixes Used
study_name_stu_svy_w1_clean.csv	Clean student survey file, wave 1	Student	300	25	stu_id, tch_id	w1_ = wave 1 s_ = student self-report
study_name_stu_svy_w2_clean.csv	Clean student survey file, wave 2	Student	300	26	stu_id, tch_id	w2_ = wave 2 s_ = student self-report
study_name_stu_svy_w3_clean.csv	Clean student survey file, wave 3	Student	298	26	stu_id, tch_id	w3_ = wave 3 s_ = student self-report
study_name_stu_records_clean.csv	Clean student records from school district, collected in wave 3	Student	300	15	stu_id, tch_id	w3_ = wave 3 d_ = district reported

* wave 1 = fall of study year; wave 2 = winter of study year; wave 3 = spring of study year

FIGURE 8.13
Example README for conveying information about files in a directory.

Templates and Resources	
Source	**Resource**
Crystal Lewis	README template for sharing information about a set of files in a directory[22]
Crystal Lewis	README template for sharing project-level information[23]

8.3.2 Changelog

When to create: Version phase

After a dataset has been collected, cleaned, and finalized, it is not uncommon to revise that file again at a later point due to errors found, or the addition of new data. However, rather than saving over previous files, it is important to use version control. Version control is a method of recording changes to a file in a way that allows you to track revision history and revert back to previous versions of a file as needed (Briney 2015; The Turing Way Community 2022). While there are automatic ways to track updates to your files through version control programs such as Git, this may not always fit into an education research workflow. Institution-approved storage locations, such as Box or SharePoint, also often have versioning capabilities. These programs save copies of your files at different points in time, allowing you to go back to previous versions. However, unless users are able to add contextual messages about changes made when saving versions (e.g., a commit message with Git), users will want to manually version their files.

Manual version control involves a two-step process. First, add a version indicator to file names (see Section 9.3 for ideas on how to version a file name). When a file is revised, a copy is saved and the indicator is updated. Second, a description of the change is recorded in a changelog—a historical record of all major file changes (UK Data Service 2023; Wilson et al. 2017) (see Figure 8.14).

Templates and Resources	
Source	**Resource**
Crystal Lewis	Changelog template[24]

Manually versioning file names and keeping an up to date changelog serves many purposes. First, it supports reproducibility. If a file is used for analysis but then that file is saved over with a new version, the original findings from that analysis can no longer be reproduced (The Turing Way Community 2022). Version control also reduces data rot (Henry 2021) by

Original file name	w1_stu_svy_clean_v01.csv
Original syntax name	w1_stu_svy_cleaning_v01.R
Description	Wave 1 clean student survey file

File version	Date created	Change	Syntax version
v01	2022-03-21	Cleaned data using original export: w1_stu_svy_raw_v01.csv	v01
v02	2022-04-11	Three students added to the raw data. Data re-cleaned using: w1_stu_svy_raw_v02.csv	v01
v03	2022-04-15	Corrected error found in recoding of `stu_gender`. Data re-cleaned using: w1_stu_svy_raw_v02.csv	v02

FIGURE 8.14
Example changelog for a clean student survey data file.

providing data lineage, allowing a user to understand where the data originated as well as all transformations made to the data. Last, it supports data confidence, allowing a user to understand what version of the data they are currently using, and to decide if they should be using a more recently created version of the file.

In its simplest form a changelog should contain the following (Schmitt and Burchinal 2011):

- The file name
- The date the file was created
- A description of the dataset (including what changes were made compared to the previous version)

It can also be helpful to record additional information such as who made the change and a link to any code used to transform the data (CESSDA Training Team 2017).

8.3.3 Data Cleaning Plan

When to create: Documentation phase

A data cleaning plan is a written proposal outlining how you plan to transform your raw data into clean, usable data. This document contains no code and is not technical skills dependent. A data cleaning plan is created for each dataset listed in your data sources catalog (see Section 8.2.2). Since this document lays out your intended transformations for each raw dataset, it allows any team member to provide feedback on the data cleaning process.

This document can be started in the documentation phase, but will most likely continue to be updated throughout the study. Typically, the person responsible for cleaning the data will write the data cleaning plans, but the documents can then be brought to a planning meeting allowing your DMWG to provide input on the plan. This ensures that everyone agrees on the transformations to be performed. Once finalized, this data cleaning plan serves as a guide in the cleaning process. In addition to the changelog, this data cleaning plan (as well as any syntax used) provides all documentation necessary to assess data provenance, a historical record of a data file's journey.

A data cleaning plan should be based on agreed-upon norms for what constitutes a clean dataset to help ensure that all datasets are cleaned and formatted consistently (see Section 14.2). These norms can be operationalized into a checklist of transformations that can inform your data cleaning plan, along with your data dictionary and other relevant documentation. We will review what types of transformations you should consider adding to your data cleaning plan in Section 14.3.

An example of a simple data cleaning plan for a student survey file.

```
1. Import raw data
2. Review data (rows and columns)
3. Remove duplicate cases if they exist (using rules from
applicable SOP and README)
4. De-identify data (bring in study IDs and remove names)
5. Rename variables based on data dictionary
6. Fix variable types as needed
7. Reverse code anx1, anx2, anx3
8. Calculate anx_mean
9. Merge in treatment from roster file
10. Add missing value codes
11. Add variable and value labels based on data
dictionary
12. Run final data validation checks
13. Export data as an SPSS file
```

8.4 Variable-Level

When we think about data management, I think this is most likely the first type of documentation that pops into people's minds. Variable-level documentation tells us all pertinent information about the variables in our datasets: Variable names, descriptions, types, and allowable values. While this documentation is often used to interpret existing datasets, it can also serve many other vital purposes including guiding the construction of data collection instruments, assisting in data cleaning, or validating the accuracy of data. We will discuss this more throughout the chapters in this book.

8.4.1 Data Dictionary

> *When to create: Documentation phase, sometimes the data capture phase for external datasets*

A data dictionary is a rectangular formatted collection of names, definitions, and attributes about variables in a dataset (Gonzales, Carson, and Holmes 2022; UC Merced Library 2023). This document is most useful if created during the documentation phase and used throughout a study for both planning and interpretation purposes (see Figure 8.15) (Lewis 2022a; Van Bochove, Alper, and Gu 2023).

A data dictionary is typically structured so that each row corresponds to a variable in your dataset, and each column represents a field of information about that variable (Broman and Woo 2018; Grynoch 2024). There are several necessary fields to include in a data dictionary, as well as several optional fields (see Table 8.1).

You should build one data dictionary for each instrument you plan to collect, including both original data collection instruments and external data sources (e.g., student education records). If there are five data sources in your data sources catalog, you should end up with five data dictionaries.

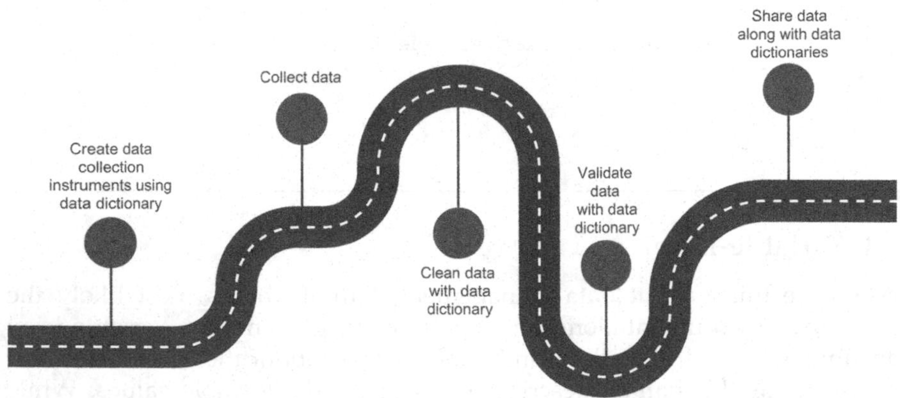

FIGURE 8.15
The many uses for a data dictionary.

TABLE 8.1

Fields to Include in a Data Dictionary

Fields to Include	Optional Fields to Include
Variable name	Skip patterns
Variable label (what is this item)	Required item (were participants allowed to skip this item)
Variable type/format	Variable universe (who received this item)
Allowable values/range (including labels associated with categorical codes)	Notes (such as versions/changes to the variable)
Assigned missing values	Associated scale/subscale
Transformations	Time periods this item is available (if study is longitudinal)
Variable origin (primary/derived)	Item order
	Remove item (should this item be removed before publicly sharing data)

8.4.1.1 Creating a Data Dictionary for an Original Data Source

When you are collecting original data, there are a few things that are helpful to have when creating your data dictionaries:

1. Your data sources catalog
   - This document (see Section 8.2.2) will provide you an overview of all the data sources you plan to collect for your project, including what measures make up each instrument.
2. Your style guide
   - We will talk more about style guides in Chapter 9, but this document will provide team or project standards for naming variables and coding response values.
3. Documentation for your measures
   - You will need a draft of all items that will be included in your instrument.
     - If you are collecting data using existing measures (i.e., existing scales, existing standardized assessments), you will want to collect any documentation on those measures, such as technical documents or copies of instruments.
       - You will want your documentation to provide information such as:
         - What items are included in the measure? What is the exact wording of the items?
         - How are items divided by subscales?

  - How are items coded? What are allowable values?
  - Are there any calculations/scoring/reverse coding needed?
  - If items are entered into a scoring program and then exported, what variables are exported?
- See Figure 8.16 for an example of the information that could be pulled from a publication if you were using the Math Anxiety Scale for Teachers (MAST) (Ganley et al. 2019).

- If you have used any items in former projects, also pull data dictionaries from previous projects as it can be helpful to collect data consistently across projects.

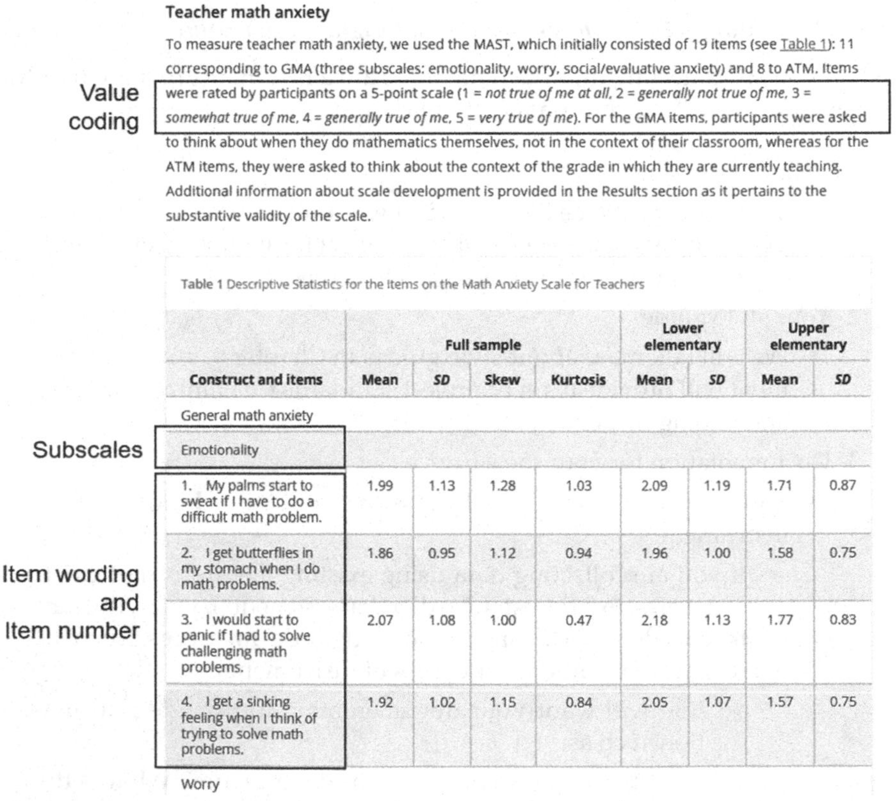

Teacher math anxiety

To measure teacher math anxiety, we used the MAST, which initially consisted of 19 items (see Table 1): 11 corresponding to GMA (three subscales: emotionality, worry, social/evaluative anxiety) and 8 to ATM. Items were rated by participants on a 5-point scale (1 = *not true of me at all*, 2 = *generally not true of me*, 3 = *somewhat true of me*, 4 = *generally true of me*, 5 = *very true of me*). For the GMA items, participants were asked to think about when they do mathematics themselves, not in the context of their classroom, whereas for the ATM items, they were asked to think about the context of the grade in which they are currently teaching. Additional information about scale development is provided in the Results section as it pertains to the substantive validity of the scale.

Value coding

Table 1 Descriptive Statistics for the Items on the Math Anxiety Scale for Teachers

	Full sample				Lower elementary		Upper elementary	
Construct and items	Mean	*SD*	Skew	Kurtosis	Mean	*SD*	Mean	*SD*
General math anxiety								
Emotionality								
1. My palms start to sweat if I have to do a difficult math problem.	1.99	1.13	1.28	1.03	2.09	1.19	1.71	0.87
2. I get butterflies in my stomach when I do math problems.	1.86	0.95	1.12	0.94	1.96	1.00	1.58	0.75
3. I would start to panic if I had to solve challenging math problems.	2.07	1.08	1.00	0.47	2.18	1.13	1.77	0.83
4. I get a sinking feeling when I think of trying to solve math problems.	1.92	1.02	1.15	0.84	2.05	1.07	1.57	0.75
Worry								

Subscales

Item wording and Item number

FIGURE 8.16
Pulling relevant information for the Math Anxiety Scale for Teachers (MAST).

4. Any data element standards that you plan to use (UC Merced Library 2023)
    - See Section 11.2.1 for an overview of existing data element standards.

All measures/items for each instrument will be included in the data dictionary. As you build your data dictionaries, consider the following:

- Item names
    - Are your variable names meeting the requirements laid out in your style guide?
    - If you have used an item before, how was it named in the previous project?
    - Are there any field standards that dictate how an item should be named?
- Item wording
    - If your items come from an existing scale, does the item wording match the wording in the scale documentation? Do you plan to reword the item?
    - If you have used an item before, how was it worded in the previous project?
    - Are there any field standards that dictate how an item should be worded?
- Item value codes for categorical items
    - If your items come from an existing scale, does your value coding (the numeric values assigned to your categorical response options) align with the coding laid out in the scale documentation?
    - If your items do not come from an existing scale, does your value coding align with the requirements in your style guide?
    - If you have used an item before, how was it coded in the previous project?
    - Are there any field standards that dictate how an item's values should be coded?
- What additional items will make up your final dataset? Consider items that will be derived, collected through metadata, or added in. All of these should be included in your data dictionary.
    - Unique study identifiers (see Section 3.3.1)
        - Primary keys (e.g., `stu_id`, `rater_id`)
        - Foreign keys as needed (e.g., `tch_id` in a student file)
    - Grouping variables (e.g., `treatment`, `cohort`, `cluster`, `block`)

    - Time component (e.g., `wave`, `time`, `year`, `session`)
    - Derived variables
        - This includes both variables your team derives (e.g., `mean scores`, `reverse coded variables`, `variable checks`) as well as variables derived from any scoring programs (e.g., `percentile ranks`, `standard scores`)
            - When describing how these variables will be calculated, make sure to account for anomalies such as calculating scores with missing data
    - Metadata (Variables that your tool collects such as `IP Address`, `completion`, `language`)
- What items should be removed before public data sharing (i.e., personally identifiable information)?

Figure 8.17 provides an example of simplified teacher survey data dictionary, including items from the MAST scale (Ganley et al. 2019).

The last step of creating your data dictionary is to review the document with your team. Gather your DMWG and review the following types of questions:

- Is everyone in agreement about how variables are named, acceptable variable ranges, how values are coded, and the variable types and formats?
- Is everyone in agreement about which participants receive each item (e.g., everyone or only a specific group)?
- Does the team want to adjust any of the item wordings?
- Does the data dictionary include everything the team plans to collect? Are any items missing?
    - If additional items are added to instruments at later points in a longitudinal collection, adding fields to your data dictionary, such as `time periods available`, can be really helpful to future users in understanding why some items may be missing data at certain time points.
- Does the dictionary include all the variables you plan to derive? (i.e., either derived by your team, or by an external scoring system)?

8.4.1.2 Creating a Data Dictionary from an Existing Data Source

Not all research study data will be gathered through original data collection methods. You may be capturing supplemental external data sources from organizations like school districts or state departments of education. If at all possible, start gathering information about your external data sources early

<table>
<tr><th>scale</th><th>subscale</th><th>var_name</th><th>origin</th><th>label</th><th>values</th><th>missing_values</th><th>type</th><th>transformations</th></tr>
<tr><td>NA</td><td>NA</td><td>tch_id</td><td>primary</td><td>Teacher unique identifier</td><td>200 – 300</td><td>NA</td><td>character</td><td>NA</td></tr>
<tr><td>NA</td><td>NA</td><td>t_svy_date</td><td>primary</td><td>Date of survey</td><td>2022-04-01 – 2022-05-20</td><td>NA</td><td>date (YYYY-MM-DD)</td><td>NA</td></tr>
<tr><td>mast</td><td>general math anxiety - emotionality</td><td>t_mast1</td><td>primary</td><td>My palms start to sweat if I have to do a difficult math problem.</td><td>1 = not true of me at all |
2 = generally not true of me |
3 = somewhat true of me |
4 = generally true of me |
5 = very true of me</td><td>-99 = missing</td><td>numeric</td><td>NA</td></tr>
<tr><td>mast</td><td>general math anxiety - emotionality</td><td>t_mast2</td><td>primary</td><td>I get butterflies in my stomach when I do math problems.</td><td>1 = not true of me at all |
2 = generally not true of me |
3 = somewhat true of me |
4 = generally true of me |
5 = very true of me</td><td>-99 = missing</td><td>numeric</td><td>NA</td></tr>
<tr><td>mast</td><td>general math anxiety - emotionality</td><td>t_mast3</td><td>primary</td><td>I would start to panic if I had to solve challenging math problems.</td><td>1 = not true of me at all |
2 = generally not true of me |
3 = somewhat true of me |
4 = generally true of me |
5 = very true of me</td><td>-99 = missing</td><td>numeric</td><td>NA</td></tr>
<tr><td>mast</td><td>general math anxiety - emotionality</td><td>t_mast4</td><td>primary</td><td>I get a sinking feeling when I think of trying to solve math problems.</td><td>1 = not true of me at all |
2 = generally not true of me |
3 = somewhat true of me |
4 = generally true of me |
5 = very true of me</td><td>-99 = missing</td><td>numeric</td><td>NA</td></tr>
<tr><td>mast</td><td>general math anxiety - emotionality</td><td>t_mast_emo_mean</td><td>derived</td><td>Mast emotionality mean score</td><td>4 – 20</td><td>-99 = missing</td><td>numeric</td><td>mean (t_mast1, t_mast2, t_mast3, t_mast4)

*not scored if >1 items are missing</td></tr>
</table>

FIGURE 8.17
Example teacher survey data dictionary.

on, during the planning and documentation phases, and begin adding that information into a data dictionary. Starting this process early will help you prepare for future data capture and cleaning processes.

- If the data source is public, you may be able to easily find codebooks or data dictionaries for the data source. If not, download a sample of the data to learn what variables exist in the source and how they are formatted.
- If the data source is non-public, request documentation ahead of time from the provider (see Section 12.3).

However, it is possible that you may not be able to access this information until you acquire the actual data during the data capture phase. If documentation is provided along with the data, begin reviewing the data to ensure that the documentation matches what you see in the data. Integrate that information into your project data dictionary.

If documentation is not provided it is important to review the data and begin collecting questions that will allow you to build your data dictionary.

1. What do these variables represent?
   - What was the wording of these items?
2. Who were the items collected from (e.g., was it only students in a specific grade level)?
3. What do these values represent?
   - Am I seeing the full range of values/categorical options for each item? Or was the range larger than what I am seeing?
   - Do I have values in my data that don't make sense for an item (e.g., a 999 or 0 in an age variable)?
4. What data types are the items currently? What types should they be?

In most situations these questions will not be easily answered without documentation and may require further detective work.

- Contact the person who originally collected the data to learn more about the instrument or data collection process.
- Contact the person who cleaned the data (if cleaned) to see what transformations they completed on the raw data.
- If applicable, request access to the original instruments to review exact question wording, item response options, skip patterns, etc.

Ultimately you should end up with a data dictionary structured similarly to Figure 8.17. You may add additional fields that help you keep track of further changes (e.g., a column for the old variable name and a column for your new variable name), and your transformations section may become more verbose as the values assigned previously may not align with the values you prefer based on your style guide or other standards. Otherwise, the data dictionary should still be constructed in the same manner mentioned in Section 8.4.1.1.

8.4.1.3 Time Well Spent

The process described in this section is a manual, time-consuming process. This is intentional. Building your data dictionary is an information seeking journey where you take time to understand your dataset, standardize items, and plan for data transformations. Spending time manually creating this document before collecting data prevents many potential errors and time lost fixing data in the future. While there are absolutely ways you can automate the creation of a data dictionary using an existing dataset, the only time I can imagine that being useful is when you have a clean dataset that you have confidently already verified is accurate and ready to be shared. However, as mentioned before, a data dictionary is so much more than a document to be shared alongside a public dataset. It is a tool for guiding many other processes in your research life cycle.

Templates and Resources	
Source	**Resource**
Crystal Lewis	Data dictionary template[25]

8.4.2 Codebook

When to create: Data cleaning and validation phase, prepare for archive phase

Codebooks provide descriptive, variable-level information as well as univariate summary statistics which allows users to understand the contents of a dataset without ever opening it. Unlike a data dictionary, a codebook is created **after** your data is collected and cleaned, and its value lies in data interpretation and data validation.

The codebook contains some information that overlaps with a data dictionary, but is more of a summary document of what actually exists in your dataset (ICPSR 2011) (see Table 8.2).

In addition to being an excellent resource for users to review your data without ever opening the file (University of Iowa Libraries 2024), this document may also help you catch errors in your data if out of range or unexpected values appear.

TABLE 8.2

Codebook Content That Overlaps and Is Unique to a Data Dictionary

Overlapping Information	New Information
Variable name	Existing values/ranges
Variable label	Existing missing values
Variable type/format	Summary statistics
Value labels	Weighting

You can create separate codebooks per dataset or have them all contained in one document, linked through a table of contents. Unlike a data dictionary, which I recommend creating manually, a codebook should be created through an automated process. Automating codebooks will not only save you tons of time, but it will also reduce errors that are made in manual entry. You can use many tools to create codebooks, including point and click statistical programs such as SPSS, or with a little programming knowledge you can more flexibly design codebooks using programs like R or SAS. The R programming language in particular has many packages that will create and export codebooks in a variety of formats from your existing dataset. Figure 8.18 is an example codebook created in R using the `memisc` package[26] (Elff 2023).

Last, you may notice as you review codebooks created by other researchers, many start with several pages of text, usually containing information about the project. When it comes time to share their data, it's common for people to combine information from their research protocol or README files, into their codebooks, rather than sharing separate documents.

Templates and Resources	
Source	**Resource**
ICPSR	Guide to Codebooks[27]
National Center for Health Statistics	Example codebook[28]

8.5 Repository Metadata

When to create: Share and archive phase

When it comes time to deposit a project or resource in a repository, you will submit two types of documentation, human-readable documentation, which includes any of the documents we've previously discussed, and metadata.

```
==========================================================================================

   t_mast1 'My palms start to sweat if I have to do a difficult math problem'

------------------------------------------------------------------------------------------

   Storage mode: double
   Measurement: ordinal
   Missing values: -99

   Values and labels                        N Valid Total

   -99 M 'missing'                          1        5.9
     1   'not at all true of me'            8  50.0  47.1
     2   'generally not true of me'         1   6.2   5.9
     3   'somewhat true of me'              4  25.0  23.5
     4   'generally true of me'             1   6.2   5.9
     5   'very true of me'                  2  12.5  11.8

==========================================================================================

   t_mast2 'I get butterflies in my stomach when I have to do math problems'

------------------------------------------------------------------------------------------

   Storage mode: double
   Measurement: ordinal
   Missing values: -99

   Values and labels                        N Percent

   1   'not at all true of me'              1     5.9
   2   'generally not true of me'           5    29.4
   3   'somewhat true of me'                3    17.6
   4   'generally true of me'               5    29.4
   5   'very true of me'                    3    17.6
```

FIGURE 8.18

Example teacher survey codebook content.

Metadata, data about your data, is documentation that is meant to be processed by machines and serves the purpose of making your files searchable (CESSDA Training Team 2017; Danish National Forum for Research Data Management 2023). Metadata aids in the cataloging, citing, discovering, and retrieving of data and its creation is a critical step in creating FAIR data (see Section 2.4.1) (Loganneider 2021; UK Data Service 2023; Wilkinson et al. 2016).

For the most part, no additional work is needed to create metadata when depositing your data in a repository. It will simply be created as part of the depositing process (CESSDA Training Team 2017; University of Iowa Libraries 2023). As you deposit your data, the repository may have you fill out a form. The details collected from this intake form are converted into both human- and machine-readable information (i.e., metadata). Figure 8.19 is an example of a metadata form for OSF (https://osf.io).

These metadata forms typically contain descriptive (description of project and files), administrative (licensing and ownership as well as

Metadata

Description

Contributors

Crystal Lewis

Resource Information

Resource type:

Resource language:

Funding/Support Information

Affiliated institutions

Date created

December 4, 2023

Date modified

December 4, 2023

Tags

No tags

Add a tag to enhance discoverability

FIGURE 8.19
Example metadata intake form for OSF.

technical information), and structural (relationships between objects) metadata (Cofield 2023; Danish National Forum for Research Data Management 2023). Commonly collected metadata elements (Dahdul 2023; Hayslett 2022) are shown in Table 8.3.

Depending on the repository, at minimum, you will enter basic project-level metadata similar to what is shown in Table 8.3. However, you may be required, or have the option, to enter more comprehensive information, such as project-level information covered in your research protocol (e.g., setting and sample, study procedures). You may also have the option to enter additional levels of metadata that will help make each level more searchable, such as file-level or variable-level metadata (Gilmore, Kennedy, and Adolph 2018; ICPSR 2023a; LDbase 2023). All the information needed for this metadata can be gathered from the documents we've discussed earlier in this chapter.

There are other ways metadata can be gathered as well. For instance, for variable-level metadata, rather than having users input metadata, repositories may create metadata from the deposited statistical data files that contain embedded metadata (such as variable types or labels) or from deposited documentation such as data dictionaries or codebooks (ICPSR 2023a).

If your repository provides limited forms for metadata entry, you can also choose to increase the searchability of your files by creating your own machine-readable documents. There are several tools to help users create machine-readable codebooks and data dictionaries that will be findable

TABLE 8.3
Common Metadata Elements

Element	Description
Title	Name of the resource
Creator	Names and institutions of the people who created the resource
Date	Key dates associated with the resource
Description	Description of the resource
Keyword/subject	Keywords describing the content of the resource
Type	Nature of your resource
Language	Language of the resource
Identifier	Unique identification code, such as a Digital Object Identifier (DOI), assigned to the resource, usually generated by the repository
Coverage	Geographic coverage
Funding agencies	Organization who funded the research
Access restrictions	Where and how your resource can be accessed by other researchers
Copyright	Copyright date and type
Format	What format is your resource in

through search engines such as Google Dataset Search (Arslan 2019; Buchanan et al. 2021; USGS 2021).

8.5.1 Metadata Standards

Metadata standards, typically field-specific, establish a common way to describe your data which improves data interoperability as well as the ability of users to find, understand, and use data. Metadata standards can be applied in several ways (Bolam 2022; Cofield 2023).

1. Formats: What machine-readable format should metadata be in (e.g., XML, JSON)?
2. Schema: What elements are recommended verses mandatory for project-, dataset-, and variable-level metadata?
3. Controlled vocabularies: A controlled list of terms used to index and retrieve data.

Many fields have chosen metadata standards to adhere to (Berry 2022; Bolam 2022) (see Figure 8.20). Some fields, like psychology (Kline et al. 2018), are developing their own metadata standards, including formats, schemas, and vocabularies grounded in the FAIR principles and the Schema.org schema (Schema.org 2023). Yet, the Institute of Education Sciences recognizes that there are currently no agreed-upon metadata standards in the field of education (Institute of Education Sciences 2023a).

Discipline	Metadata standard
General	Dublin Core (DC) Metadata Object Description Schema (MODS) Metadata Encoding and Transmission Standard (METS) DataCite Metadata Schema
Arts and Humanities	Categories for the Description of Works of Art (CDWA) Visual Resources Association (VRA Core) Text Encoding Initiative Guidelines (TEI)
Astronomy	Astronomy Visualization Metadata (AVM)
Biology	Darwin Core
Ecology	Ecological Metadata Language (EML)
Geographic	Content Standard for Digital Geospatial Metadata (CSDGM)
Social sciences	Data Documentation Initiative (DDI)

FIGURE 8.20
A sampling of field metadata standards.

If you plan to archive your data, first check with your repository to see if they follow any standards. For example, both the OSF (Gueguen 2023) and Figshare (Figshare 2023) repositories currently use the DataCite schema, while ICPSR uses the DDI standard (ICPSR 2023a). If the repository does use certain standards, work with them to ensure your metadata adheres to those standards. Some repositories may even provide curation support free or for a fee. But as I mentioned earlier, depending on your repository, adding metadata to your project may require no additional work on your part. The repository may simply have you enter information into a form and convert all information for you.

If no standards are provided by your repository and you plan to create your own metadata, you can choose any standard that works for you. Oftentimes researchers may choose to pick a more general standard such as DataCite or Dublin Core (University of Iowa Libraries 2024), and in the field of education, most researchers are at least familiar with the DDI standard so that is another good option. Remember, if you do choose to adhere to a standard, this decision should be documented in your data management plan.

Resources

Source	Resource
Registry of Research Data Repositories	A searchable database of repositories and their metadata standards[29]

8.6 Wrapping It Up

At this point your head might be spinning from the number of documents we've covered. It's important to understand that while each document discussed provides a unique and meaningful purpose, you don't have to create every document listed. In data management we walk a fine line between creating sufficient documentation and spending all of our working hours perfecting and documenting every detail of our project. Choose the documents that help you record and structure your processes in the best way for your project while also giving yourself grace to stop when the documents are "good enough". Each document you create that is well organized and well maintained will improve your data management workflow, decrease errors, and enhance your understanding of your data.

Notes

1 https://docs.google.com/document/d/1LqGdtHg0dMbj9lsCnC1QOoWzIsnSNRTSek6i3Kls2Ik
2 https://docs.google.com/spreadsheets/d/1kn4A0nR4loUOSDn9Qysd3MqFJ9cGU91dCDM6x9aga-8
3 https://www.rsb.org.uk/images/biologist/2020/Apr_May_2020_67-2/67.2_Handbook.pdf
4 https://osf.io/mdh87/wiki/home/
5 https://www.notion.so/help/guides/how-to-build-a-wiki-for-your-company
6 https://eur-synclab.github.io/
7 https://osf.io/yzsc8
8 https://osf.io/h4kwf
9 https://osf.io/6rgfd
10 https://eur-synclab.github.io/data-management/data-security.html#data-security-protocol
11 https://uofnelincoln.sharepoint.com/:b:/s/ResearchComplianceServicesSharepoint/Ebr9awmFio1Mj3PTgb-GnSIBoS7xSua7uT-jePT2qtGTlw?e=RbRs50.
12 https://style.tidyverse.org/
13 https://hwpi.harvard.edu/files/sdp/files/sdp-toolkit-coding-style-guide.pdf.
14 https://osf.io/5n2ke
15 https://osf.io/3yp7x
16 https://osf.io/q6g8d
17 https://figshare.com/articles/preprint/IRB_Protocol_Template/13218797
18 http://orrp.osu.edu/files/2011/10/GuidelinesforWritingaResearchProtocol.pdf
19 https://docs.research.missouri.edu/human_subjects/templates/Social_Behavioral_Educational_Protocol_Template.docx
20 https://depts.washington.edu/wildfire/resources/protckl.pdf

21 https://osf.io/yq3np
22 https://osf.io/tk4cb
23 https://osf.io/d3pum
24 https://osf.io/trw6b
25 https://osf.io/ynqcu
26 https://osf.io/d3tx4
27 https://www.icpsr.umich.edu/files/deposit/Guide-to-Codebooks_v1.pdf
28 https://ftp.cdc.gov/pub/Health_Statistics/NCHS/Dataset_Documentation/NHIS/2020/adult-codebook.pdf
29 https://www.re3data.org/

9

Style Guide

A style guide provides general rules for the formatting of information. As mentioned in Chapter 8, style guides can be created to standardize procedures such as variable naming, variable value coding, file naming, file structure, and even coding practices.

Style guides create standardization within and across projects. The benefits of using them consistently include:

- Creating interoperability: This allows data to easily be combined or compared across forms or time.
- Improving interpretation: Consistent and clear structure, naming, and coding allow your files and variables to be findable and understandable to both humans and computers. This in turn prevents errors such as accidentally using the wrong file or incorrectly interpreting a variable.
- Increasing reproducibility: If the organization of your file paths, file naming, or variable naming constantly change, it undermines the reproducibility of any data management or analysis code you have written.

Style guides can be created for individual projects, but they can also be created at the team level, to be applied across all projects. Most importantly, they should be created before a project kicks off so you can implement them as soon as your project begins. If you do not have a team-wide style guide already created, you most likely will want to create a project-level style guide during your planning phase so that you can begin setting up your directory structures and file naming standards before you start creating and saving project-related files.

Style guides can be housed in one large document, with a table of contents used to reference each section, or they can be created as separate documents. Either way, style guides should be stored in a central location that is easily accessible to all team members (such as a team or project wiki), and all team members should be trained, and periodically retrained, on the style guide to ensure adherence to the rules. If all team members are not consistently implementing the style guide, then the benefits of the guide are lost (Figure 9.1).

DOI: 10.1201/9781032622835-9

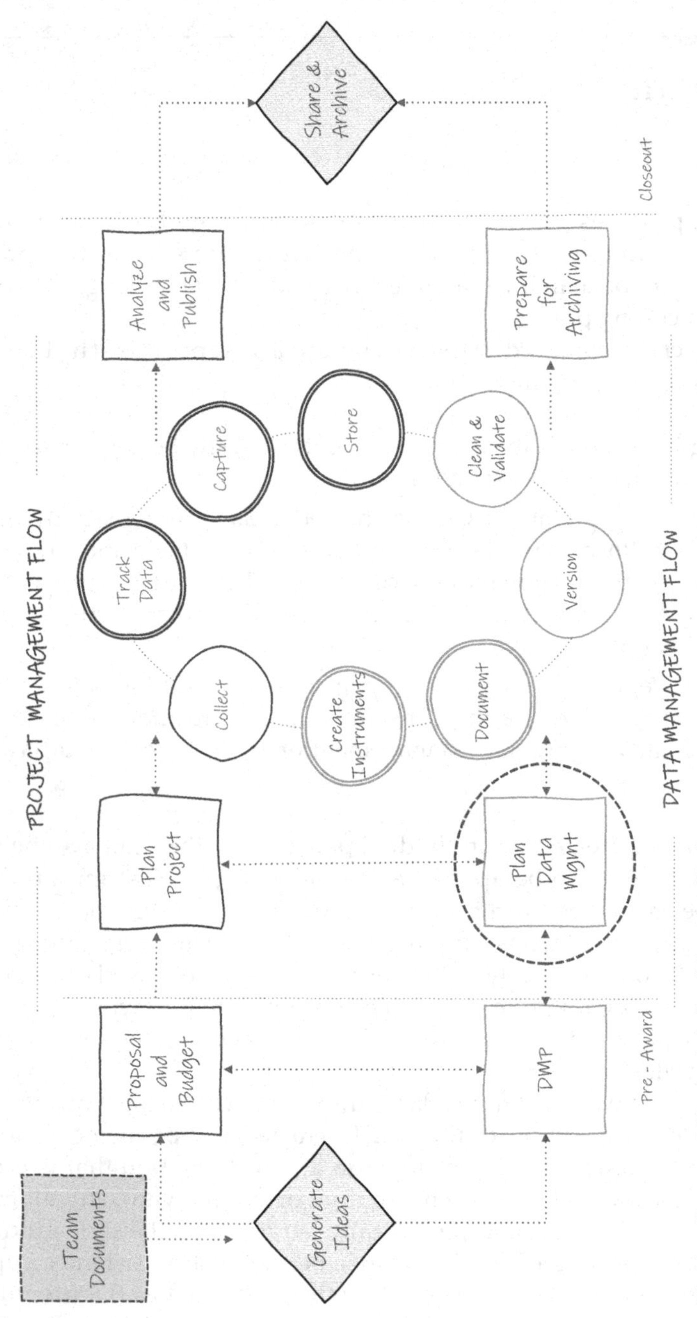

FIGURE 9.1
Style guide development in the research project life cycle.

For the remainder of this chapter, we will spend time reviewing best practices to consider when creating style guides for the following purposes:

1. Structuring directories
2. Naming files
3. Naming variables
4. Assigning variable values
5. Writing code

9.1 General Good Practices

Before we dive into particular types of style guides, there are a few things to know about how computers read names in order to understand the "why" behind some of these practices.

1. Avoid spaces.
   - Command line operations and some operating systems do not support them, so it is best to avoid them all together. Furthermore, they can often break a URL when shared.
   - The underscore (_) and hyphen (-) are generally good delimiters to use in place of spaces. An exception to this rule is denoted in Section 9.4.
2. With the exception of (_) and (-), avoid special characters.
   - Examples include but are not limited to ?, ., *, \, /, +, ', &, ' '.
   - Computers assign specific meaning to many of these special characters.
3. There are several existing naming conventions that you can choose to add to your style guide. Different naming conventions may work better for different purposes. Using these conventions help you to be consistent with both delimiters and capitalization, which not only makes your names more human-readable but also allows your computer to read and search names easier.
   - Pascal case (ScaleSum)
   - Snake case (scale_sum)
   - Camel case (scaleSum)
   - Kebab case (scale-sum)
   - Train case (Scale-Sum)

4. Character length matters. Computers are unable to read names that surpass a certain character length. This applies to file paths, file names, and variable names. Considerations for each type of limit are reviewed below.

9.2 Directory Structure

When deciding how to structure your project directories (the organization of folders and files within an operating system), there are several things you want to consider.

When structuring your folders:

- First, consider organizing your directory into a hierarchical folder structure to clearly delineate segments of your projects and improve searchability.
    - The alternative to using a folder structure is using metadata and tagging to organize and search for files (Cakici 2017; Fuchs and Kuusniemi 2018; Krishna 2018).
- When creating your folder structure, strike a balance between a deep and shallow structure.
    - Too shallow leads to too many files in one folder which is difficult to sort through.
    - Too deep leads to too many clicks to get to one file, plus file paths can max out with too many characters. A file path includes the full length of both folders and file name.
        - An example file path with 73 characters `W:\team\project_new\data\wave1\student\survey\pn_w1_stu_svy_clean_v02.csv`
    - Examples of file path limits:
        - SharePoint/OneDrive path limit is 400 characters (Microsoft 2023)
        - Windows path limit is 260 characters (Ashcraft 2022)
- Create folders that are specific enough that you can limit access.
    - For example, you will want to limit user access to folders that hold personally identifiable information (PII).
    - To protect any files that you don't want others to accidentally edit (for example your clean datasets), also consider making some files "read only".

- Decide if you want an "archive" folder to move old files into or if you want to leave previous versions in the same folder.
- Ultimately, create a structure that is consistently applied.
    - For example, if your project includes three phases of an intervention, consider creating "phase" folders at the top of your directory with content folders under each **or** creating content folders at the top of your directory and with phase folders in each. Do not mix and match methods. This leads to confusion when searching for files.

Top-level phase folders

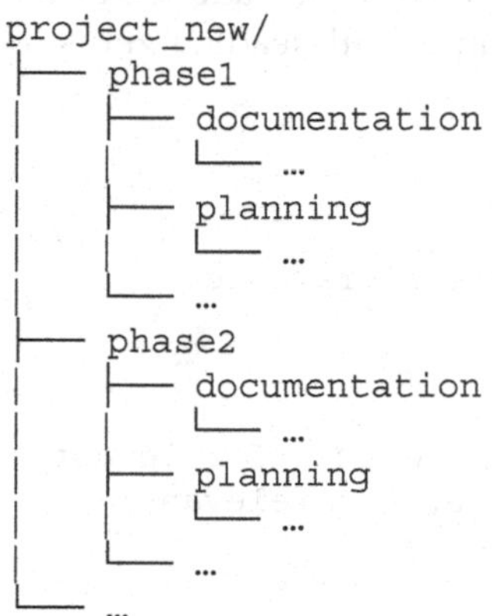

Top-level content folders

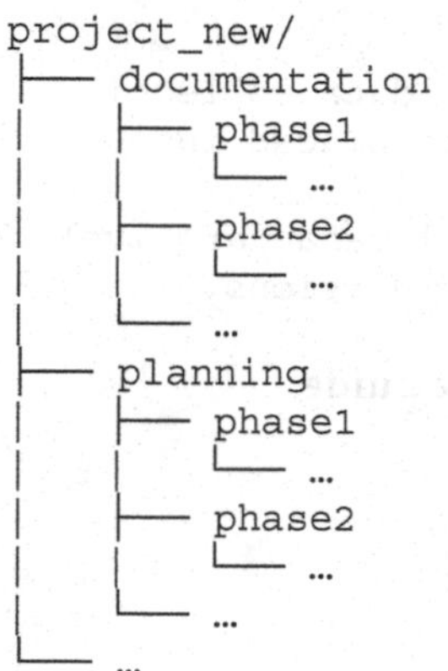

When naming your folders:

- Consider setting a character limit on folder names (again to reduce problems with hitting path character limits).
- Make your folder names meaningful and easy to interpret.
- Don't use spaces in your folder names.
    - Use (_) or (-) to separate words.
- With the exception of (-) and (_), don't use special characters in your folder names.
- Be consistent with delimiters and capitalization. Follow an existing naming convention (as mentioned in Section 9.1).
- If you prefer your folders to appear in a specific order, add the order number to the beginning of the folder name, with leading zeros to ensure proper sorting (`01_`, `02_`).

Example directory structure style guide

```
1. All project directories follow this hierarchical
metadata structure
    - Level 1: Name of project
    - Level 2: Life cycle folders
    - Level 3: Data collection wave folders (if relevant)
    - Level 4: Participant group folder (if relevant)
    - Level 5: Specific content folder/s
    - Level 6: Archive folder
2. All folders should be named according to these rules
    - Meaningful name but no longer than 20 characters
    - No spaces or special characters except (_)
    - Only use lower case letters
    - Use (_) to separate words
    - Consistently named across waves of data collection
3. All previous versions of files must be placed into
their respective "archive" folder
    - A changelog should be placed in all data "archive"
folders to document changes between file versions
```

Example directory structure created using a style guide

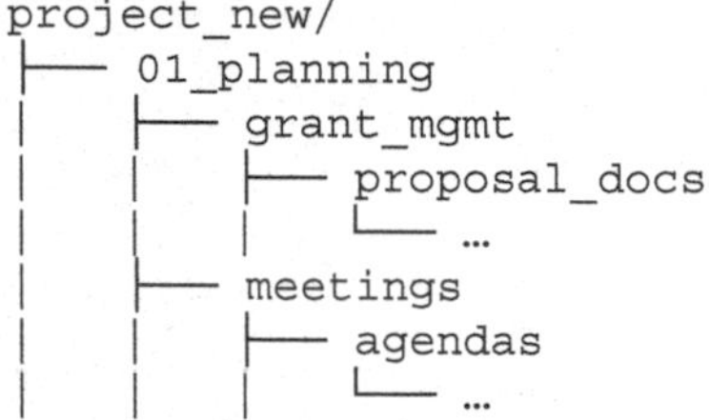

```
project_new/
├── 01_planning
│   ├── grant_mgmt
│   │   ├── proposal_docs
│   │   │   └── ...
│   ├── meetings
│   │   ├── agendas
│   │   │   └── ...
```

```
│     └── ...
├── 02_documentation
│     ├── consent_forms
│     │     └── ...
│     ├── data_dictionaries
│     │     └── ...
│     ├── data_sources_catalog.xlsx
│     ├── instrument_copies
│     │     └── ...
│     ├── research_protocol.docx
│     ├── sops
│     │     └── ...
│     └── ...
├── 03_data_collection
│     ├── hiring_materials
│     │     └── ...
│     ├── scheduling_materials
│     │     └── ...
│     └── ...
├── 04_tracking
│     ├── parent_consents
│     │     └── ...
│     └── participant_tracking_database.accdb
├── 05_data
│     ├── cohort1
│     │     ├── student
│     │     │     ├── survey
│     │     │     │     ├── clean
│     │     │     │     │     ├── archive
│     │     │     │     │     │     └── changelog.txt
│     │     │     │     ├── raw
│     │     │     │     │     ├── archive
│     │     │     │     │     │     └── changelog.txt
│     │     │     │     └── ...
│     │     │     └── ...
│     │     └── ...
│     └── ...
└── ...
```

9.3 File Naming

As seen in Figure 9.2, naming files in a consistent and usable way is hard. We are often in a rush to save our files and maybe don't consider how unclear our file names will be for future users (including ourselves).

FIGURE 9.2
"Documents", from xkcd.com.

Our file names alone should be able to answer questions such as:

- What are these documents?
- When were these documents created?
- Which document is the most recent version?

A file naming style guide helps us to name files in a way that allows us to answer these questions. You can have one overarching file naming guide, or you may have file naming guides for different purposes that need different organizational strategies (e.g., one naming guide for project meeting notes, another naming guide for project data files). Let's walk through several conventions to consider when naming your files.

- Make names descriptive (a user should be able to understand the contents of the file without opening it).
- No PII should be used in a file name (e.g., participant name).

- Never use spaces between words.
  - Use (`-`) or (`_`) to separate words.
  - It is worth noting that underscores may be difficult to read when file paths are shared in links that are underlined to denote that the path is clickable (for example when sharing a SharePoint link to a document).
- With the exception of (`_`) and (`-`), never use special characters.
- Be consistent with delimiters and capitalization. Follow an existing naming convention (see Section 9.1).
- Consider limiting the number of allowable characters to prevent hitting your path limit.
  - For instance, Harvard Longwood Research Data Management (2023) recommends keeping file names to 40-50 characters.
- Format dates consistently and do not use forward slashes (/) to separate parts of a date. It is beneficial to format dates using the ISO 8601 standard in one of these two ways (International Organization for Standardization 2017):
  - *YYYY-MM-DD* or *YYYYMMDD*
    - Using the ISO 8601 standard ensures that dates are consistently formatted and correctly interpreted (e.g., "06-01-2023" is interpreted as DD-MM-YYYY in Europe, while it is often interpreted as MM-DD-YYYY in the U.S.).
  - While using delimiters between parts of the date adds characters to your variable names, it also may be clearer for users to interpret. Either of these date formats will be sortable.
- When manually versioning file names (see Section 8.3.2), pick a consistent indicator to use.
  - One method is to add a number to the file name. Using this method, consider left padding single numbers with a 0 to keep the file name the same length as it grows (`v01`, `v02`).
  - Another method is to add a date to the file name, using the ISO 8601 standard.
- If your files need to be run in a sequential order, add the order number to the beginning of the file name, with leading zeros to ensure proper sorting (`01_`, `02_`).
- Choose abbreviations to use for common phrases (student = `stu`).
  - This helps reduce file name character lengths and also creates standardized, searchable metadata, which can allow you to more easily, programmatically retrieve files (for example, retrieve all files containing the phrase "stu_obs_raw").

- Keep redundant metadata (information) in the file name.
  - This reduces confusion if you ever move a file to a different folder or send a file to a collaborator. It also makes your files searchable.
  - For example, always put the data collection wave in a file name, even if the file is currently housed in a specific wave folder, always put the project acronym in the file name, even if the file is currently housed in a project folder, or always put the word "raw" or "clean" in a data file name, even if the file is housed in a "raw" or "clean" folder.
- Choose an order for file name metadata (e.g., project -> time -> participant -> instrument).

Example file naming style guide

```
1. Never use spaces between words
2. Never use special characters except (_) or (-)
3. Use (_) to separate words
4. Only use lower case letters
5. Keep names under 50 characters
6. Format dates as YYYY-MM-DD
7. Use the following metadata file naming order
  - Order of use (if relevant - add a 0 before single
digits)
   - Project
   - Cohort/Wave (if relevant)
   - Participant group
   - Instrument
   - Further description
   - Version
     - For raw data files, use date
     - For clean data files, code files, or SOPs use
version number (add a 0 before single digits)
8. Use the following abbreviations
   - student = stu
   - teacher = tch
   - survey = svy
   - observation = obs
   - wave = w
   - project new = pn
```

Example file names created using a style guide

```
pn_stu_svy_sop_v03.docx
pn_w1_tch_obs_raw_2023-11-03.csv
pn_w1_stu_svy_raw_entry1_2023-11-15.csv
pn_w1_stu_svy_raw_entry2_2023-11-15.csv
```

```
pn_w1_stu_svy_clean_syntax_v01.R
pn_w1_stu_svy_clean_syntax_v02.R
```

9.4 Variable Naming

This style guide will be a necessary document to have before you start to create your data dictionaries. Below are several considerations to review before developing your variable naming style guide. These are broken into two types of rules, those that are nonnegotiable requirements that really should be included in your style guide (if you do not follow these rules you will run into serious problems in interpretation for both humans and machines), and then best practice suggestions that are recommended but not required.

Mandatory:

- Don't name a variable any keywords or functions used in any programming language (such as `if`, `for`, `repeat`) (R Core Team 2023; Stangroom 2019).
- Set a character limit.
  - Most statistical programs have a limit on variable name characters.
    - SPSS is 64
    - Stata is 32
    - SAS is 32
    - Mplus is 8
    - R is 10,000
  - With this said, do not limit yourself to 8 characters based on the fact that one future user may use a program like Mplus. Consider the balance between character limit and interpretation. It is very difficult to make good human-readable variable names under 8 characters. It is much easier to make them under 32. Also, most of your users will be using a program with a limit of 32 or more. If you have one potential Mplus user, they can always rename your variables for their specific analysis.
- Don't use spaces or special characters, except (_). They are not allowed in most programs.
  - Even the (-) is not allowed in programs such as R and SPSS as it can be mistaken for a minus sign.
  - While (.) is allowed in R and SPSS it is not allowed in Stata so it's best to avoid using it.

- Do not start a variable name with a number. This is not allowed in many statistical programs.
- Do not embed indicator information into variable names.
    - Cohort, treatment, site, or other grouping information should be added as its own variable in a dataset, not embedded into variable names. This allows your data to be combined across groups as needed.
- All variable names should be unique.
    - This absolutely applies to variables within the same dataset, but it should also apply to all variables across datasets within a project (e.g., across a teacher survey and a student survey). The reason is, at some point you may merge data across forms and end up with identical variable names (which programs will not allow).
    - The exception to this rule is if you are collecting the same variables across time. In this case, identical variables should be named consistently across waves of data collection to allow easier comparison of information.
        - If an item is named `anx1` in the fall, name that same item `anx1` again in the spring (see Section 9.4.1 for a discussion on accounting for time).
- If you substantively change an item (substantive wording OR response options change) after at least one round of data has been collected, version your variable names to reduce errors in interpretation.
    - For example, revised `anx1` becomes `anx1_v2`.

Suggested:

- Names should be meaningful.
    - Instead of naming gender `q1`, name it `gender`.
    - If a variable is a part of a scale, consider using an abbreviation of that scale plus the scale item number (`anx1`, `anx2`, `anx3`).
        - Not only does this allow you to easily associate an item with a scale, but it also allows you to programmatically select and manipulate scale items (for example, sum all items that start with "anx").
- If you have used an item before, consider keeping the variable name the same across projects. This can be very useful if you ever want to combine data across projects. It also allows you to easily reuse code snippets across projects (e.g., for scoring a measure).

  - If you choose not to reuse variable names across projects, it will be important for your team to create a document mapping comparable names to facilitate potential future data integration.
- Choose standard abbreviations to denote the types of variables you are working with (standard score = `ss`). Using controlled vocabularies improves interpretation and makes data exploration and manipulation easier (Riederer 2020).
- Be consistent with delimiters and capitalization. Follow an existing naming convention. Most programming languages are case-sensitive, so consider this when choosing a convention that is feasible for your workflow.
  - Snake case (scale_sum) – preferred method for variable names
    - While pascal case and camel case are also options, the use of underscores helps more clearly delineate relevant pieces of metadata in your variable names.
  - Kebab case (scale-sum) – don't use for variable names
  - Train case (Scale-Sum) – don't use for variable names
- If a variable includes a "select all" option, start all associated variables with the same prefix (`cert_elem`, `cert_secondary`, `cert_leader`, `cert_other`). Again, this allows you to easily see grouped items, as well as easily programmatically manipulate those items as needed.
- Consider denoting reverse coding in the variable name to reduce confusion (`anx1_r`).
- Include an indication of the reporter in the variable name (student self-report = `s`, district student records = `d`, teacher self-report = `t`, parent report = `p`).
  - Doing this allows you to consistently name comparable variables to improve interpretation while still creating unique variable names.
    - For example, if you collect student gender from a survey and from district student records you can use the names `s_gender` and `d_gender`
- Choose an order for variable name metadata (e.g., reporter -> variable name -> item number/type).

Example variable naming style guide

```
1. Use snake case
2. Keep names under 32 characters
3. Use meaningful variable names
```

```
4. Use unique variable names within and across data
sources
5. Use consistent variable names across waves of data
collection
6. If part of a scale, use scale/subscale abbreviation
plus item number from the scale.
 - If the scale has been used in another project, keep
the same name from previous projects.
7. Include an indication of the reporter as a prefix in
the variable name
  - student self-report = s_
  - teacher self-report = t_
  - district student records = d_
8. Denote reverse coded variables using suffix _r
9. Follow this metadata order for variable names
  - Reporter
  - Item name (or scale abbreviation)
  - Subscale abbreviation (if relevant)
  - Item number (if relevant)
  - Variable type (if relevant)
  - Recode (if relevant)
9. Use the following abbreviations
  - mean = m
  - standard score = ss
  - percentile rank = pr
  - other, open text = text
  - date = dt
```

Example variable names created using a style guide

```
s_wj_lwi_dt
s_wj_lwi1
s_wj_lwi2
s_wj_lwi_ss
s_wj_lwi_pr
s_gender
d_gender
t_profdev
t_profdev_text
t_stress5
t_stress5_r
t_stress_m
```

9.4.1 Time

Before moving on, there is one last consideration for variable names. If your data is longitudinal (i.e., you are collecting repeated measures), you may need to add rules for accounting for time in your variable names as well.

Recall from Section 3.3.2, there are two ways you can link and structure longitudinal data, in wide format or long format.

1. If combining data over time in long format, no changes need to be made to your variable names. Variable names should be identically named over time. To account for a time component, you will simply include a new variable (e.g., `time`, `year`, `wave`) and add the appropriate value for each row.
2. If combining data in wide format, you will need to concatenate time to all of your time-varying variable names (i.e., not your subject unique identifier). This removes the problem of having non-unique variable names (e.g., `anx1` in wave 1 and `anx1` in wave 2) and allows you to interpret when each variable was collected. How you concatenate time to your variable names is up to you. Just make sure to continue adding time consistently to all variable names (i.e., same location, same format) and remember to follow variable naming best practices (e.g., never start a variable name with a number).

Before adding a time component, either as a new variable or as part of a variable name, it's important to decide what values you want to assign to time. This will depend entirely on your study design and how you intend to use time in your analyses. When working with cohorts, it can be helpful to choose generic time values that allow you to combine samples collected in the same relative time periods. For example:

- `wave 1` = fall of the study year
- `wave 2` = spring of the study year
- `wave 3` = fall of the follow up year

However, when not working with cohorts or if you have a dataset that does not fall within your predefined data collection periods, you can choose any values that work for you. Figure 9.3 provides just a few examples of how you might account for time in your data based on different scenarios.

With all of that said, **during an active project, it is actually best to not add a time component to your data**, and to store each dataset as a distinct file, with a clear file name that denotes the appropriate time period (e.g., `pn_w1_stu_svy_clean_v01.csv`). There are a few benefits of this method.

1. Naming variables consistently over time (with no time component added) allows you to easily reuse your data collection and data capture tools, as well as your cleaning code, each wave (Reynolds, Schatschneider, and Logan 2022).

Scenario	When collected	Time component added as a variable	Time component concatenated to variable names
A pre/post survey	Fall 2023 and Spring 2024	Variable name: time Values: pre, post	pre_varname post_varname
A survey collected over three time periods, for two cohorts of participants	Cohort 1 Fall 2023, Spring 2024, Fall 2024 Cohort 2 Fall 2024, Spring 2025, Fall 2025	Variable name: wave Values: 1, 2, 3	w1_varname w2_varname w3_varname
Student records demographic data collected for two school years	2022-23 and 2023-24	Variable name: sch_year Values: 23, 24	varname_23 varname_24

FIGURE 9.3
Examples of adding time to your data based on a variety of scenarios.

2. Storing files separately prevents you from potentially wasting time combining your data in a way that ends up not actually being useful or from wasting time merging files that later need to be re-combined because you find an error in a dataset at some point.

Therefore, add rules to your variable naming style guide around how to concatenate time to your variable names, but make an asterisk saying that this time component should not be added until it is necessary (e.g., when you need to combine files, when you are publicly sharing data). Once you are ready, it is fairly easy to add a time variable or to concatenate time to variable names using a statistical program, such as R,[1] or even in a program like Microsoft Excel.

9.5 Value Coding

Oftentimes in education research we codify categorical values. This coding of values helps in data entry, data scoring, and data analysis. As an example, rather than referring to the lengthier values of "yes" or "no" in a variable, we may code those values into a code/label pair. The code can be numeric (e.g., "yes" = 1 | "no" = 0) or character (e.g., "yes" = 'y' | "no" = 'n'), depending on your needs. Ultimately, only the code appears in your data, while the code/label pair is represented in your data dictionary, allowing users to interpret the meaning of each code.

If you are planning to code any categorical variable values for your study, it can be helpful to include general guidelines in your style guide for assigning those codes. Some general good practices are outlined below.

First, if you are using a pre-existing measure, assign codes and labels in the manner that the technical documentation tells you to assign codes. That will be important for any further derivations you need to make later based on those measures. Similarly, if you have used items before in previous projects, consider keeping the same coding from prior projects to improve interpretation and usability of those variables.

Otherwise, if you are assigning new codes, follow these guidelines to improve interpretation and decrease user errors:

- Codes must be unique.
  - Do: Assign "yes" = 1 | "no" = 0
  - Don't: Assign "yes" = 1 | "no" = 1
- Codes must be consistent within a variable.
  - Do: For `gender` assign "male" = 'm'
  - Don't: For `gender` allow "male" = 'm' or 'M' or 'Male' or 'male'
- Codes must be consistent across time.
  - Do: For `anx1` assign "yes" = 1 | "no" = 0 in wave 1 **and** wave 2
  - Don't: For `anx1` assign "yes" = 1 | "no" = 0 in wave 1 **but** "yes" = 1 | "no" = 2 in wave 2
- Codes should be consistent across the project.
  - Do: Assign "yes" = 1 | "no" = 0 as the value for all yes/no items
  - Don't: Assign "yes" = 1 | "no" = 0 for some variables, and "yes" = 1 | "no" = 2 for others
    - The exception here is if a pre-existing measure determines how values are coded. In that case, there may be some inconsistency across items.
- Align codes with response options as best as possible
  - Do: Assign "none" = 0 | "1" = 1 | "2" = 2 | "3 or more" = 3
  - Don't: Assign "none" = 1 | "1" = 2 | "2" = 3 | "3 or more" = 4
- Likert-type scale codes should be logically ordered
  - Do: Assign "strongly disagree" = 1 | "disagree" = 2 | "agree" = 3 | "strongly agree" = 4
  - Don't: Assign "strongly disagree" = 1 | "disagree" = 3 | "agree" = 4 | "strongly agree" = 2
  - The exception here is if a pre-existing measure tells you to code variables in a different way.

9.5.1 Missing Value Coding

There is little agreement about how missing data should be assigned (White et al. 2013). There are essentially two options.

1. You can choose to leave all missing values blank.
    - Benefits of this option is that there is no chance of assigned missing value codes (e.g., -999) being mistaken as actual values.
    - The concern with this method is that there is no way to discern if the value is truly missing, or was potentially erased by accident or skipped over during data entry (Broman and Woo 2018).
    - Also, some statistical programs do not allow blank values (e.g., Mplus), and therefore missing values will need to be assigned at some point. Yet, as I mentioned earlier in this chapter, it is best to not make decisions based on one potential use case. It is better to make decisions based on what is the most reasonable way to assign missing values for a general audience.
2. The other option is to define missing codes and add them to your data. This code can be numeric (e.g., "missing" = -999) or character (e.g., "missing" = 'NA') and it may be one consistent code applied to all missing data, or it may be multiple codes assigned for different types of missing data.
    - One benefit of this method is that this removes the uncertainty that we had with blank cells. If a value is filled, we are now certain the value was not deleted or skipped over during data entry.
    - Another benefit is that this allows you to specify distinct reasons for missing data (e.g., "Not Applicable" = -97, "Skipped" = -98) if that is important for your study.
    - The biggest problem that can occur with this method is that either your codes could be mistaken for actual values (if someone misses the documentation on missing values), or if you use a value that does not match your variable type, then you introduce new variable type issues (e.g., if 'NULL' is used in a numeric variable, that variable will no longer be numeric)

Ultimately, whichever method you choose, there are several guidelines you should follow.

1. If you decide to fill with defined missing codes, use values that match your variable type (e.g., numeric codes for numeric variables) (ICPSR 2020; White et al. 2013)

TABLE 9.1

Example Missing Value Code Schema Used for Numeric Variables

Code/Value	Label
-99	Unit nonresponse (entire instrument not completed)
-98	Item skipped
-97	Item not applicable
-96	Don't know

- There is, however, some merit to using text to define missing values in numeric variables to prevent incorrect use of missing values. If you try to run a mean on your variable, you will be immediately notified that this is not possible because your variable will be stored as a character column. If you do not care about the different types of missingness, you could easily then choose to change all missing codes to blank. However, if you do care about the types of missingness and want to keep that included in your variable, you will need to match the variable type.

2. If you use numeric values, use extreme values that do not actually occur in your data.
3. Use your values consistently within and across variables.

In your value coding style guide, you can add general rules to follow, or it may be an appropriate place to actually designate a missing value coding schema for your project (see Table 9.1).

> **NOTE**
>
> It is important to note here that there is a difference between a string value of "NA" or "NULL" that is inserted, versus an `NA` or `NULL` value that is assigned to blank values by a specific tool. For instance, blank numeric values in R are represented as the symbol `NA`. Yet this value is treated as missing, not as the string "NA".

9.6 Coding

If your team plans to clean data using code it can be very helpful to create a coding style guide. This style guide can be tailored to a specific language that all staff will use (e.g., R or Stata), or it can be written more generically to apply to any coding language staff use to clean data. Below is a small sampling of good coding practices to consider adding to your guide. If you are looking for guides for a specific language, it can be very helpful to search online for existing style guides in that language.

- Consider building and implementing coding templates (Castañeda 2019; Farewell 2018).
  - Templates can standardize the format of syntax files (such as using standard headers to break up code).
  - They also standardize the summary information provided at the beginning of your syntax (e.g., code author, project name, date created) (Alexander 2023).

One example of what a template might look like in R.

```
#### Overview ####
# Associated project:
# Script purpose:
# Data cleaning plan followed:
# Created by:
# Date created:
# Code checked by:
# Code checked date:


#### Workspace setup ####

# Settings, packages, root paths


#### Data import ####


#### Cleaning code section 1 ####


#### Cleaning code section 1 ####
...
```

- Use comments throughout your code to clearly explain the purpose of each code chunk.
  - While your syntax may seem intuitive to you, it is not necessarily clear to others. As you write your code, comment every step in your syntax, explaining what that specific line of code is doing. The format of these comments will depend on your coding language.
    - R uses # at the start of a comment
    - SPSS and Stata use * at the start of a comment
- Improve code readability by using (San Martin, Rodriguez-Ramirez, and Suzuki 2023; Wickham 2021)
  - spaces

- indentation
- setting a line limit for your code (e.g., 80 characters)

- Use relative file paths for reproducibility.
    - In a point and click environment (e.g., Microsoft Excel), we typically open or read in a file by going to `file -> open` and navigating to the file's location. However, when writing code, we import a file by writing a file path, a location where a file lives, in our syntax. Rather than writing the full, absolute file path, it is a good practice to write the path relative to the directory you are working in (Wickham, Çetinkaya-Rundel, and Grolemund 2023). Setting absolute file paths in syntax reduces reproducibility because future users may have different file paths.

```
# Example absolute file path
"/Users/crystal/project_new/data/raw/pn_stu_svy_clean_
v01.csv"

# Example relative file path
"raw/pn_stu_svy_clean_v01.csv"
```

- If you create objects in your program (like you do in R or Python), consider adding object naming rules similar to variable naming rules.
    - No spaces in object names
    - No special characters except (_) to separate words
    - No names that are existing program keywords (`if`, `for`, etc.)
- Don't repeat yourself.
    - Reduce duplication, improve efficiency, and increase your ability to troubleshoot errors by following the DRY (don't repeat yourself) principle. Consider using functions, loops, or macros for repetitive code chunks.
- Record session information for future users.
    - Information about software/package versions and operating systems used should be recorded in a text or log file to increase the reproducibility of your code. If users run into errors running your code, this information may help them troubleshoot.

Note

1 https://osf.io/xumg4

10

Data Tracking

During your project you will want to be able to answer both progress and summary questions about your recruitment and data collection activities.

1. How many participants consented to be in our study? How many have we lost during our study and why?
2. How much progress have we made in this cycle of data collection? How much data do we have left to collect?
3. How many forms did we collect each cycle and why are we missing data for some forms?

Questions like these will arise many times throughout your study for both internal project coordination purposes, as well as for external progress reporting and publication purposes. Yet, how will you answer these questions? Will you dig through papers, search through emails, and download in-progress data, each time you need to answer a question about the status of your project activities? A better solution is to track all project activities in a participant tracking database.

A participant tracking database is an essential component of both project management and data management. This database contains all study participants, their relevant study information, as well as tracking information about their completion of project milestones. This database has two underlying purposes.

1. To serve as a roster of study participants and a "master key" (Pacific University Oregon 2014) that houses both identifying participant information as well as assigned unique study identifiers.
2. To aid in project coordination and reporting, tracking the movement of participants as well as completion of milestones, throughout a study.

This database is considered your single source of truth concerning everything that happened throughout the duration of your project. Any time a participant consents to participate, drops from the study, changes their name, completes a data collection instrument, is provided a payment, or moves locations, a project coordinator, or other designated team member, updates the information in this one location. Tracking administrative information

DOI: 10.1201/9781032622835-10

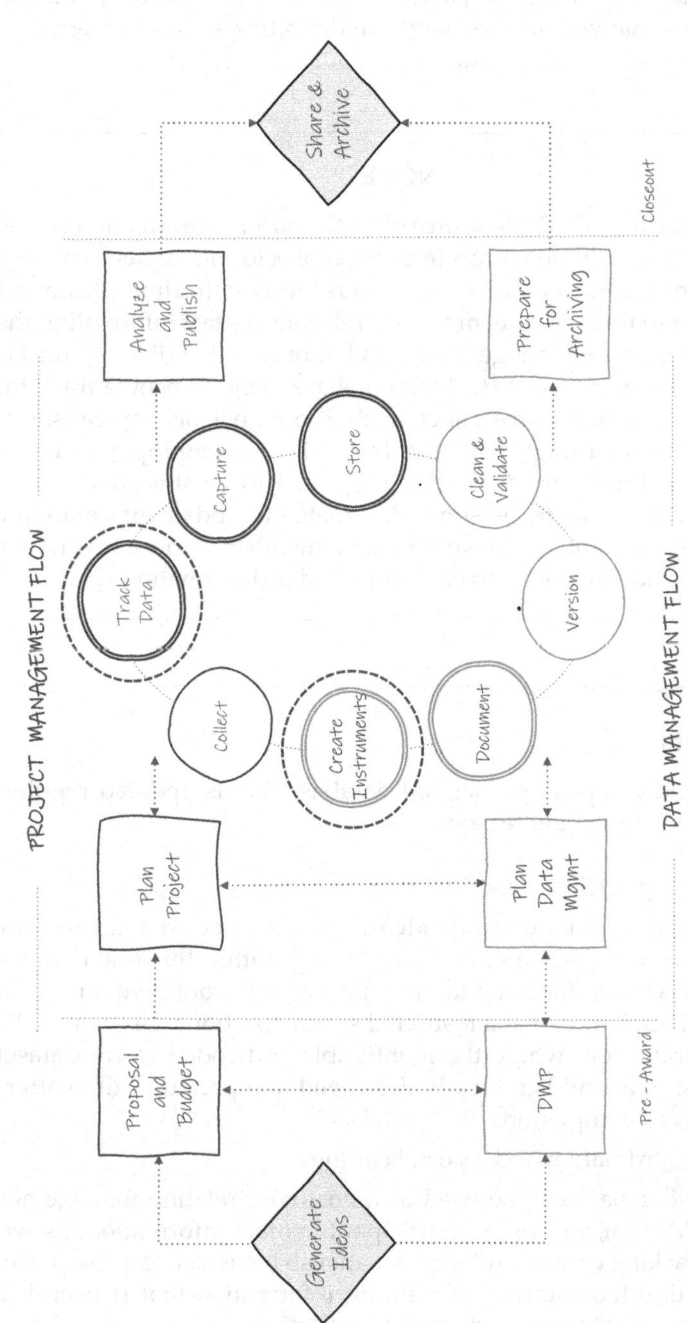

FIGURE 10.1
Tracking in the research project life cycle.

in this one database, rather than across disparate spreadsheets, emails, and papers, ensures that you always have one definitive source to refer to when seeking answers about your sample and your project activities.

NOTE

I want to reiterate this single source of truth concept. Information is often coming in from multiple sources (e.g., data collectors in the field, emails to project coordinators from teachers, conversations with administrators). It is important to train your team that all relevant contact information that is gleaned (e.g., name change, new email, moved out of district) must be updated in the participant tracking database alone. If people track this information in other sources, such as their own personal spreadsheets, there is no longer a single source of truth, there are multiple sources of truth. This makes it very difficult to keep track of what is going on in a project. Whether a single person is designated to update information in this database, or multiple, make sure team members know either how to update information or who to contact to update information.

10.1 Benefits

A thorough and complete participant database that is updated regularly is beneficial for the following reasons:

1. Protecting participant confidentiality
    - Assigning unique study identifiers (i.e., codes) that are only linked to a participant's true identity within this one database is necessary for maintaining participant confidentiality. This database is stored in a restricted secure location (see Chapter 13), separate from where the identifiable and coded study datasets are stored, and is typically destroyed at a period of time after a project's completion.
2. Project coordination and record keeping
    - This database can be used as a customer relation management (CRM) tool, storing all participant contact information, as well as tracking correspondence. It can also be used as a project coordination tool, storing scheduling information that is useful for planning activities such as data collection.

- Integrating this database into your daily workflow allows your team to easily report the status of data collection activities (e.g., as of today we have completed 124 out of 150 assessments). Furthermore, checking and tracking incoming data daily, compared to after data collection is complete, reduces the likelihood of missing data.
- Last, thorough tracking allows you to explain missing data in reports and publications (e.g., teacher 1234 went on maternity leave).

3. Sample rostering
   - At any time, you can pull a study roster from this database that accurately reflects a participant's current status. The tracking information contained in this tool also aids in the creation of documentation including the flow of participants in your CONSORT diagram (see Section 8.2.6).
4. Data cleaning
   - As part of your data cleaning process, all raw dataset sample sizes should be compared against what is reported as complete in your tracking database to ensure that no participants are missing from your final datasets.
   - Furthermore, this database can be used for de-identifying data. If data is collected with identifiers such as name, a roster from the tracking database can be used to merge in unique study identifiers so that name can be removed. A similar process can be used to merge in other assigned variables contained in the database such as treatment or cohort.

10.2 Building Your Database

While the tracking phase appears after collection in Figure 10.1, it is most beneficial to build this database before you begin recruiting participants, typically during the same time that you are building your data collection tools, in the create instruments phase. This way, as your team recruits participants, you can record information such as name, consent status, and any other necessary identifying contact information in the participant database and begin assigning participants study IDs.

While a project coordinator can build this database, it can be helpful to consult with a data manager, or someone with database expertise, when creating this system. This ensures that your system is set up efficiently and comprehensively.

This database may be a standalone structure, used only for tracking and anonymization purposes, or it may be integrated as part of your larger study system, where all study data is collected and/or entered as well.

10.2.1 Comparing Database Types

Before we discuss how to build this database, it is helpful to have a basic understanding of the benefits of using different types of databases. There are essentially two types of databases commonly used for tracking—relational and non-relational.

1. Relational database
   - Relational databases are typically built in a database software system (e.g., FileMaker, Microsoft Access). In this database, information is organized into tables made up of records (rows) and fields (columns), and tables can be related through keys (Bourgeois 2014; Chen 2022).
2. Non-relational database
   - Here I am loosely using the term "non-relational database" to describe tables of information that are not linked (think tabs in a spreadsheet). This type of database is usually built in a spreadsheet program (e.g., Microsoft Excel). While this is technically not considered a database, we will continue to use the term "database", even if this system is built using a spreadsheet program.

Your study design can inform the type of database you choose to build. While relational databases may be more involved to build, they are also more efficient to use if your study includes a variety of related entities (e.g., students and teachers), tracked over waves of time. However, if you are only tracking one group of participants (e.g., just students) for one wave of data collection, or you have a fairly small study (e.g., participant numbers < 30), then a relational database might be overkill and a simple spreadsheet system will work just fine. Figure 10.2 is a flow chart showing the kinds of decisions to consider when choosing which type of database to build. This decision tree should only be used to help guide your discussion; it does not contain hard-and-fast rules. Rules such as N > 30 should be replaced with criteria that make sense for your project and team.

In the following sections we will review the benefits of relational databases compared to non-relational databases, especially when working with more complex studies.

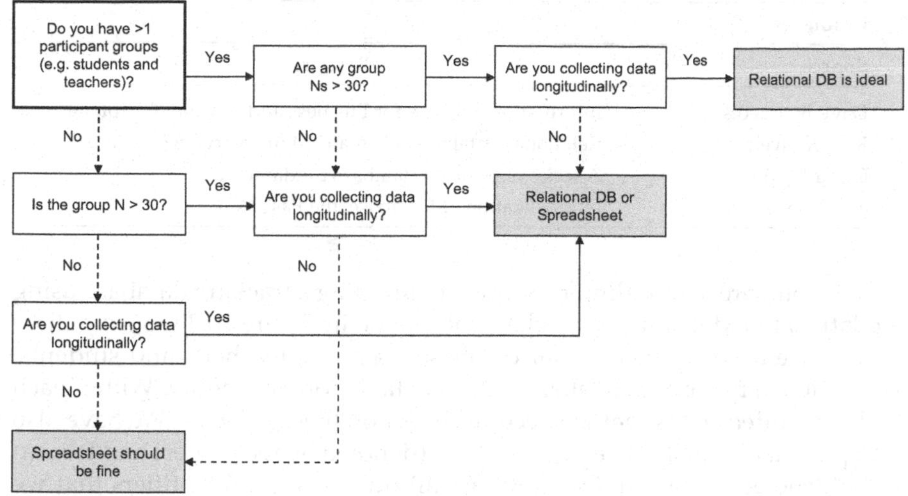

FIGURE 10.2
A decision tree for choosing whether to build a participant tracking database using a relational database or a spreadsheet.

10.2.1.1 Relational Database

Using a relational database to track participant information, compared to disconnected tabs in a spreadsheet, has many benefits including reducing data entry errors and improving efficiency (Borer et al. 2009).

There are three general steps for building a relational database.

1. Create tables made up of fields (i.e., variables)
2. Choose one or more primary key fields to uniquely identify rows in those tables. These keys should not change at any point. Typically these keys are your assigned unique study IDs.
3. Create relationships between tables through both primary and foreign keys (see Section 3.3.1 for a refresher on primary and foreign keys)

Understanding all the ways to optimize your relational database is outside the scope of this book and also not always necessary for our purposes. Here, the most important thing to consider when building a relational database is to **not duplicate information across tables**. Any one field should only need to be updated in one location, never more than one. If you want to learn more about building databases, there are many freely available resources.

Resources	
Source	**Resource**
Dave Bourgeois	Information Systems for Business and Beyond, Chapter 4[1]
Kim Nguyen	Relational Database Schema Design Overview[2]
Omar Elgabry	A series of posts on database fundamentals[3]
The Nobles	Normalization of Database, the Easy Way[4]

Let's compare a very simple example of building a tracking database using a relational model and a non-relational model. In Figure 10.3 we have three entities we need to track in our database—schools, teachers, and students. We built a very simple database with one table for each entity. Within each table we added fields that we need to collect on these subjects. We have also set up our tables to include primary keys (denoted by rectangles) and foreign keys (denoted by ovals). Our keys are all unique study identifiers that we have assigned to our study participants (see Section 10.4 for more information on assigning these IDs).

We can see that across each table we have no duplicated information. The Student Table only contains student-level information, the Teacher Table only contains teacher-level information, and the School Table only contains school-level information. This is a huge time saver. Imagine if a teacher's last name changes. Rather than updating that name in multiple places, we now only update it once, in the teacher table, and make a note of the previous name in the notes field.

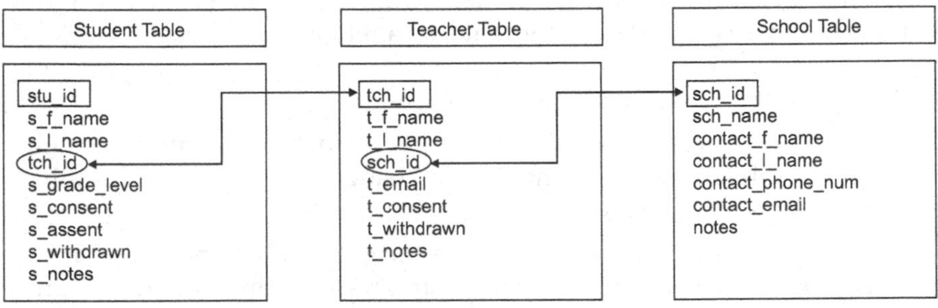

FIGURE 10.3
Participant database built using a relational model.

If we want to see a table with both student and teacher information, we can simply query our database (i.e., make a request) to create a new table. In some programs, this type of querying may be a simple point and click option, in other programs it may require someone to write some simple code that can then be used at any time by any user.

Say, for example, we needed to pull a roster of students for each teacher. We could easily create and run a query, such as this one written in SQL (Structured Query Language), that joins the student and teacher tables from Figure 10.3 by `tch_id` and then pulls the relevant teacher and student information from both tables.

```
SELECT Teacher.t_l_name, Teacher.t_f_name, Student.s_l_
 name, Student.s_f_name, Student.s_grade_level
FROM Student LEFT JOIN Teacher ON Student.tch_id =
 Teacher.tch_id
ORDER BY t_l_name, t_f_name, s_l_name, s_f_name
```

The resulting output is Table 10.1.

TABLE 10.1

Example Roster Created by Querying Our Relational Database Tables

t_l_name	t_f_name	s_l_name	s_f_name	s_grade_level
Clark	Jana	Arnold	Darnell	2
Clark	Jana	Watts	Irene	2
Ramirez	Bill	Dixon	Ernesto	4
Ramirez	Bill	Gibson	Emma	4
Ramirez	Bill	Webster	Grant	4

Depending on the design of your study and the structure of the database model, writing these queries can become more complicated. It's important to strike a balance between creating a structure that reduces inefficiencies in data entry but also creating something that isn't too complicated to work with based on the expertise of your team.

10.2.1.2 Non-Relational Database

Now imagine that we built a non-relational model, such as three tabs in an Excel spreadsheet, to track our participant information (see Figure 10.4). Since we are unable to link these tables together, we need to enter redundant information into each table (denoted by rectangles) in order to see that information within each table without having to flip back and forth across tables to find the information we need. For example, we now have to enter repeating teacher and school names in the Student Table, and if any teacher names change, we will need to update it in both the Teacher Table and in the Student Table for every student associated with that teacher. While this model may work fine for smaller studies, you can imagine how duplicating information would increase data entry time and create more opportunity for data entry errors for a larger, more complicated study with several tables of information (Borer et al. 2009).

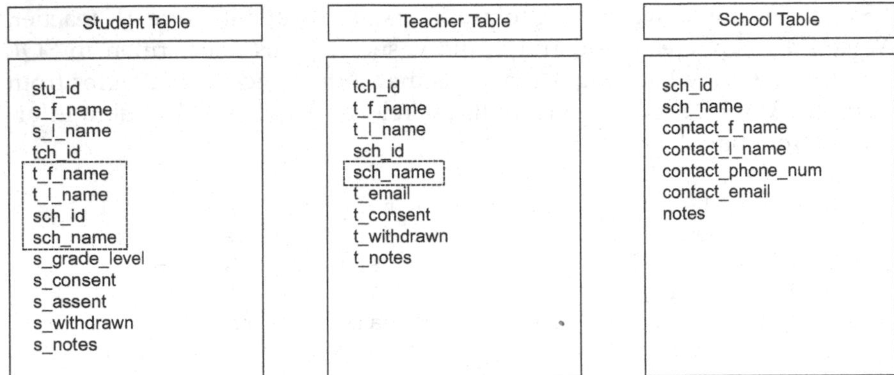

FIGURE 10.4
Participant database built using a non-relational model with duplicated variables denoted by rectangles.

10.2.2 Designing the Database

Before you can begin to design your database, you will need to think through the following pieces of information.

1. Do you want to use a relational or non-relational database?
2. How many tables do you want to construct?
   - Consider entities (e.g., student, teacher, school)
   - Consider purpose (e.g., enrollment info, wave 1 data collection tracking, wave 2 data collection tracking)
3. What fields do you want to include in each table?
4. If using a relational database, what fields will you use to relate tables?

Once you make decisions regarding these questions, you can begin to design your database. It can be helpful to visualize your database schema during this process. In Figure 10.5 I am designing a relational database schema for a scenario where I will be collecting information from teachers and schools, over two waves of data collection.

I have designed this database model in this way:

1. I have four tables total.
   - Two tables (Teacher Info and School Info tables) have information that should be fairly constant based on my project assumptions (name, email, consent, one-time documents received).
     - If at any time this information changes (e.g., withdrawn status, new last name, new contact person), I would update that

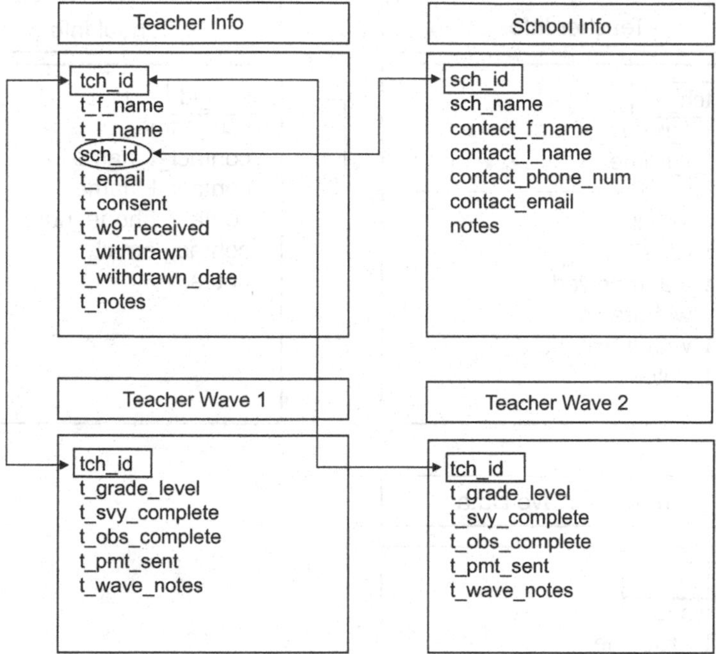

FIGURE 10.5
Example participant database relational model using two separate tables for tracking across waves.

information in the appropriate table and make a note of when and why the change occurred in my notes field.

- Two tables are for my longitudinal information.
  - This is where I will track my data collection activities each wave, as well as any information that may change each wave, again based on the assumptions of my project. For example, I may put grade level in my longitudinal tables if I collect data across years because I assume it's possible that teachers may switch grade levels.

2. I have connected my tables through primary and foreign keys (`tch_id` and `sch_id`).
3. With information separated into four tables, I can also now limit access as needed (e.g., only allow data entry staff access to the de-identified tables, or restricting entry to only the current wave of data preventing accidental overwriting of existing data).

The model in Figure 10.5 is absolutely not the only way you can design your tables. There may be more efficient or more appropriate ways to design this

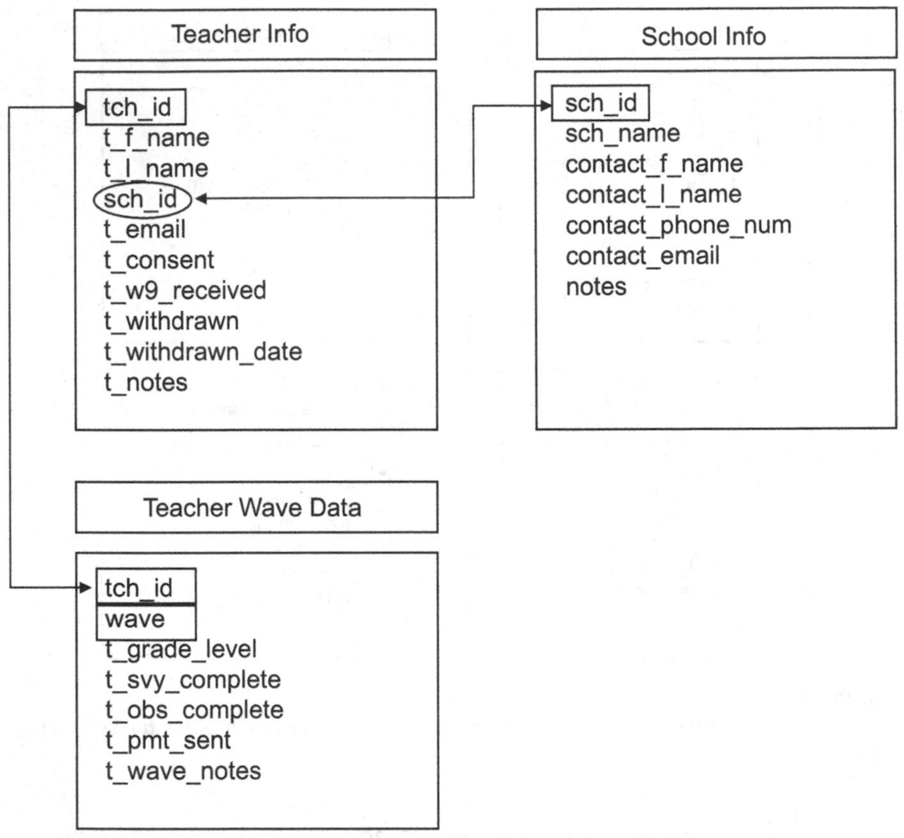

FIGURE 10.6
Example participant database relational model using one table to track data across waves.

database, but again as long as you are not duplicating information, build what works for you. As an example of a potentially more efficient way to structure this database, I could combine all waves of data collection into one table and create a concatenated (or compound) primary key that uses both `tch_id` and wave to uniquely identify rows since `tch_id` would now be duplicated for each wave of data collection (see Figure 10.6).

If we entered some data into the Teacher Wave Data table in Figure 10.6, it might look something like Figure 10.7. We can see that `tch_id` repeats but the rows are unique when combined with `wave`.

While these examples are for a fairly simple scenario, you can hopefully see how you might extrapolate this model to more entities and more waves of data collection, as well as how you might modify it to better meet the needs of your specific project.

tch_id	wave	t_grade_level	t_svy_complete	t_obs_complete	t_pmt_sent	t_wave_notes
235	1	1	complete	complete	sent	
236	1	2	complete	complete	sent	
235	2	1	incomplete	incomplete	not sent	out on leave
236	2	4	complete	complete	sent	

FIGURE 10.7
An example of data that contains a concatenated primary key.

> **NOTE**
>
> If your study involves anonymous data collection, you will no longer be able to track data associated with any specific individual. However, it is still helpful to create some form of a tracking system. Creating a simplified database, with tables based on your sites for instance (school table, district table) allows you to still track your project management and data collection efforts (e.g., number of student surveys received per school per wave, payment sent to school).

10.2.3 Choosing Fields

As you design your participant tracking database model, you will also need to choose what fields to include in each table. The fields you choose to include will be dependent on your particular study design. While your participant tracking database may be the same database you enter all your study data, for the purposes of this chapter we are only considering fields that are relevant for project coordination and participant de-identification. We are not concerned with fields that are collected as part of your data collection measures (e.g., survey items). You can consider your participant tracking database as an **internal database** that is only used for coordination, summary, and linking purposes. This is not a database where you would export data for external data sharing.

Below are ideas of fields you may consider adding to your database. Depending on the design and assumptions of your study, some of these may be collected once, others may be collected more than once, longitudinally.

Ideas of fields to collect:

- Primary keys (Study IDs)
- Foreign keys

- Names (participants and sites)
- Contact information
- Other necessary linking identifiers (double IDs, district/school IDs)
- Consent/assent status
- Inclusion/exclusion criteria status
- Enrollment status
- Withdrawn status
- Relevant dates
- Randomization (treatment/control)
- Grouping information (cohort)
- Information relevant to project coordination (grade level, class periods, block schedules)
- Summary information that may be helpful for participant flow diagrams (# of consents sent out, # of students in class, # of teachers in school)
- Administrative data status (W-9 received, MOUs received)
- Data collection status (unique fields for each instrument)
- Data collection administrator names or IDs
- Incentive status (gift cards sent out)
- Notes
    - Reasons for changes (for example changes in name, email)
    - Reasons for movement/withdraw
    - Communication with participants
    - Reasons for missing data
    - Errors in data

10.2.3.1 Structuring Fields

As you choose your fields you also need to make decisions about how you will structure those fields.

1. Set data types for your fields (e.g., character, integer, date)
    - Restrict entry values to only allowable data types to reduce errors
2. Set allowable values and ranges
    - For example, a categorical status field may only allow "complete", "partially complete" or "incomplete"

3. Do not lump separate pieces of information together in a field
    - For example, separate out first name and last name into two fields
4. Name your fields according to the variable naming rules we discussed in Chapter 9

10.2.4 Choosing a Tool

There are many criteria to consider when choosing a tool to build your database in.

- Choose a tool that is customizable to your needs.
    - Can you build a relational table structure?
    - Can you export files? Can you connect to the database via application programming interfaces (APIs)?
    - Can you query data?
- Choose a tool that is user-friendly.
    - You don't want a tool with a steep learning curve for users.
- If you are running a project across multiple sites, consider the accessibility of the tool.
    - For example, you may want a tool that is cloud-based so that all site coordinators can access it.
    - You may also want to make sure multiple users can access it at the same time.
- Choose a tool that is interoperable.
    - For instance, some tools may have difficulties running on certain operating systems.
- Consider cost and licensing.
    - What products do you already have access to (i.e., your institution has a license for)?
- Consider security.
    - Which tools are approved by your institution to protect the sensitivity level of this data (see Chapter 4)?
    - Can you limit access to the entire database? To specific tables?
        - If multiple people are entering data, you may want to restrict access/editing capabilities for some tables.
    - Protect data loss.
        - Can you back up the system?
        - Can you protect against overwriting data?

    - Can you keep versions of the database in case a mistake is ever made and you need to go back to an older version?
- Data quality protection.
    - Can you set up data quality constraints (e.g., restrict input values/types)?

There are many tool options you can choose from. A sampling of those options is below. These tools represent a wide range from the criteria in this section. Take some time to review your options to see which one best meets your needs.

- Microsoft Access
- Microsoft Excel
- Quickbase
- Airtable
- REDCap
- Claris FileMaker
- Google Sheets and Google Forms
- Forms that feed into a relational database, maintained using a SQL database engine such as SQLite, MySQL, or PostgreSQL

10.3 Entering Data

Your last consideration when building your database will be, how do you want your team to enter data into your database? There are many ways to enter data including manually entering data, importing data, integrating your data collection platform and your tracking database, or even scanning forms using QR codes. While some of those options may work great for your project, here we are going to talk about the two simplest and most common options—manually entering data into a tabular view, and manually entering data into a form.

10.3.1 Entering Data in a Tabular View

Your first option is to manually enter data in a tabular view for each participant in a row (see Figure 10.8). This would be the most common, or possibly the only option, when using spreadsheet tools such as Microsoft Excel. However, you can also use this option when entering into other database tools such as Microsoft Access or FileMaker. Depending on the tool, this stu_w1

stu_id	tch_id	svy_complete	obs_complete	notes
12406	236	complete	complete	
12412	235	complete	complete	
12419	235	complete	incomplete	Absent on observation day 2023-10-23
12428	236	complete	complete	
12431	235	partially complete	complete	Survey sent back to the field on 2023-11-08
12433	236	complete	complete	

FIGURE 10.8
Example tabular view data entry.

might have a name such as datasheet or table view. There are both pros and cons to this method.

- Pros: This is the quickest and easiest method. It also allows you to view all the data holistically.
- Cons: This method can lead to errors if someone enters data on the wrong row/record. It can also more easily lead to accidental deletion or overwriting of data.

10.3.2 Entering Data in a Form

Your second option is to create a form (also called a data entry screen) that is linked to your tables (see Figure 10.9). As you enter data in your forms, it automatically populates your tables with the information. This option is

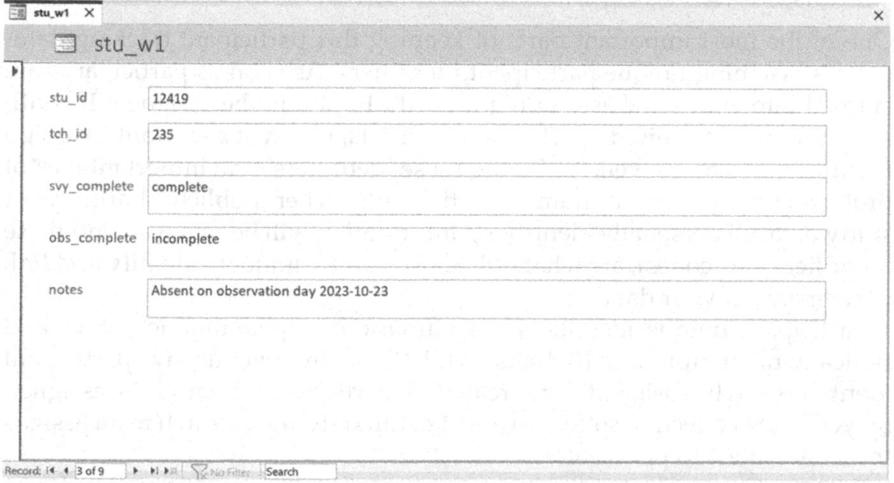
stu_w1

stu_w1

stu_id: 12419
tch_id: 235
svy_complete: complete
obs_complete: incomplete
notes: Absent on observation day 2023-10-23

Record: 3 of 9 | No Filter | Search

FIGURE 10.9
Example form view data entry.

possible in many systems including Microsoft Access, FileMaker, REDCap, and even Google Forms which populates into Google Sheets.

- Pros: This method reduces data entry errors as you are only working on one participant form at a time.
- Cons: Depending on the tool you use and your preferred layout, it may take some time, and possibly expertise, to set up the data entry forms.

NOTE

If your participant tracking database is separate from your data collection tools, all information will need to be entered by your team using one of the ways mentioned in this section. However, if your participant tracking tool is also your data collection tool (such as those who collect electronic data using REDCap), fields such as data collection status (e.g., `svy_complete`) may not need to be manually entered. Rather they may be automated to populate as "complete" once a participant submits their responses in the data collection tool.

10.4 Creating Unique Identifiers

One of the most important parts of keeping this participant tracking database is assigning unique participant identifiers. As soon as participants are entered into your database, a unique study ID should be assigned. If confidentiality was promised to schools or districts, you will also want to assign identifiers to sites as well. Assigning these identifiers is an important part of protecting the privacy of human participants. When publicly sharing your study data, all personally identifying information will be removed and these identifiers (i.e., codes), are what will allow you to uniquely identify and link participants in your data.

Participant unique identifiers are numeric or alphanumeric values and typically range from 2 to 10 digits. While there are several ways participant identifiers can be assigned (e.g., created by participants themselves, assigned by your data collection software), most commonly, the research team assigns these identifiers to participants.

Before assigning identifiers, it can be very helpful to develop an ID schema during your planning phase (see Table 10.2), and document that schema in an SOP (see Section 8.2.7).

TABLE 10.2

Example of a Study ID Schema

stu_id	tch_id	sch_id
12000–1300	200–300	40–80

This schema sets the parameters for how participant identifiers are assigned as they are entered into your tracking database (i.e., format, range). There are several best practices to consider when developing your participant ID schema.

1. Participants must keep this same identifier for the entire project.
   - Having a static participant ID allows you to track the flow of each participant through your study and can also provide the added benefit of helping to measure dosage in intervention studies.
   - Keeping the same unique identifier even applies in circumstances where a participant has the opportunity to be re-recruited into your study (as seen in Figure 10.10). In this situation, the participant still keeps the same ID and you will use a combination of variables to identify the unique instances of that participant (e.g., `stu_id` and `cohort`).
   - When you have multiple rounds of recruitment, it is important to have a procedure in place to check for participants who may already be in your database (e.g., search participant names before adding them into the database). Without this system in place, it is possible that you bring a participant back into your database under a new ID.

stu_id	cohort	grade	anx1	anx2
12412	1	4	2	3
12419	1	5	1	2
12428	1	3	4	1
12412	2	5	4	3

FIGURE 10.10

Example of keeping the same participant IDs for the entire study.

2. Participant identifiers must be unique within and across entities.
   - For example, no duplicating IDs within students or across teachers and schools.
   - Not duplicating within entities is imperative to maintain uniqueness of records, while not duplicating across reduces confusion about who a form belongs to and reduces potential errors.
3. The identifier should be randomly assigned and be completely distinct from any personal information to protect confidentiality.
   - Do not sort by identifying information (e.g., names, date of birth) and then assign IDs in sequential order.
   - Do not group by identifying information (e.g., grade level, teacher) and then assign IDs in sequential order.
   - Do not include identifying information (e.g., initials) as part of an identifier.
4. Do not embed project information into the ID if that information has the potential to change.
   - Some researchers prefer to embed a project-level ID or acronym into a participant ID to help with tracking of information, especially when running multiple studies using identical forms across studies. This is absolutely okay because it is assumed this information never changes.
   - However, embedding a time indicator, such as wave or session, into an identifier variable guarantees that your identifiers will not remain constant. This information should be added to your dataset in other ways (i.e., either as its own variable or concatenated to variable names).
   - Embedding information such as teacher IDs, school IDs, treatment, or cohort also has the potential to cause problems. In longitudinal studies, depending on the study design, it is possible that students move to other study teachers, teachers move to other study schools, or participants get re-recruited into other cohorts. Any of these issues would cause problems if this information was embedded into an ID because the ID would no longer reflect accurate information and would require IDs to be changed, breaking best practice #1. Again, this information can be tracked as separate variables (e.g., `tch_id`, `sch_id`, `cohort`, `treatment`) and added to forms and datasets as needed. Figure 10.11 is an example dataset where `stu_id` = 12428 moved into another teacher's classroom in wave 2 and this would have caused issues if teacher ID was embedded into the `stu_id` variable.

stu_id	w1_tch_id	w2_tch_id	w1_anx1	w2_anx1
12412	235	235	1	2
12419	238	238	2	4
12428	238	237	3	1

FIGURE 10.11
Example of a student changing teachers during a research study.

5. Last, while less important during the data tracking phase, in your study datasets these identifiers should be stored as character variables. Even if an ID variable is all numbers, it should be stored as character type. This helps prevent people from inappropriately working with these values (i.e., taking a mean of an ID variable).

> **NOTE**
>
> The only time you will not assign unique study identifiers is when you collect anonymous data. In this situation you will not be able to assign identifiers since you will not know who participants are. However, it is still possible to assign identifiers to known entities such as school sites if anonymity is required.

10.5 Summary

The tracking phase is one of the most important data management practices in an education research project life cycle. Anecdotally, I have seen several teams, without a participant tracking system for their project, end up with lost or unusable data and struggle to recall the details of their data collection efforts. While optimizing the design of your database and entry system is very helpful in reducing both inefficiencies and errors, don't let yourself get lost in the details. Build a system that works for your team and your project. The key takeaways to focus on during this phase are the following:

1. Build a system for tracking the intake of information, as well as for storing your participant key, and keep it up to date.
2. Keep one single source of truth. Don't have information stored in multiple locations.

3. Keep your tracking system secure. Don't allow unauthorized access to participant information (see Chapter 13 for more information).
4. Assign your participant unique IDs using best practices covered in Section 10.4.
5. Integrate data tracking into your daily workflow during data collection (see Section 11.3.3 for more information).

Notes

1 https://pressbooks.pub/bus206/chapter/chapter-4-data-and-databases/
2 https://medium.com/@kimtnguyen/relational-database-schema-design-overview-70e447ff66f9
3 https://medium.com/omarelgabrys-blog/database-introduction-part-1-4844fada1fb0
4 https://medium.com/swlh/normalization-of-database-the-easy-way-98f96a7a6863

11

Data Collection

When collecting original data as part of your study (i.e., you are administering a survey or assessment as opposed to using an externally collected data source), data management best practices should be interwoven throughout your data collection process. Unfortunately, quality data doesn't just happen because an instrument is created and data is collected. It takes careful consideration, structure, and care on the part of the entire team. The number one way to improve the integrity of your data is to spend time planning your data collection efforts. Not only does planning minimize errors, it also keeps your data secure and valid, and relieves future data cleaning headaches.

11.1 Quality Assurance and Control

When planning your data collection efforts, you'll want to have your data sources catalog available (see Section 8.2.2). This document will be a guide during your data collection planning period. Recall that each row in that document is an original instrument to be collected for your study. Some of your data sources may also include external datasets, which we will discuss in Chapter 12.

In addition to planning data collection logistics for your original data sources (i.e., how will data be collected, who will collect it, and when), teams should spend time prior to data collection anticipating potential data integrity problems that may arise during data collection and putting procedures in place that will reduce those errors (DIME Analytics 2021a; Northern Illinois University 2023). As shown in Figure 11.1, creating data collection instruments is typically a collaborative effort between the project management and data management team members. Even if the project management team builds the data collection tools, the data management team is overseeing that the data collected from the tools aligns with expectations set in the data dictionaries. In this chapter we will review two types of practices that

DOI: 10.1201/9781032622835-11

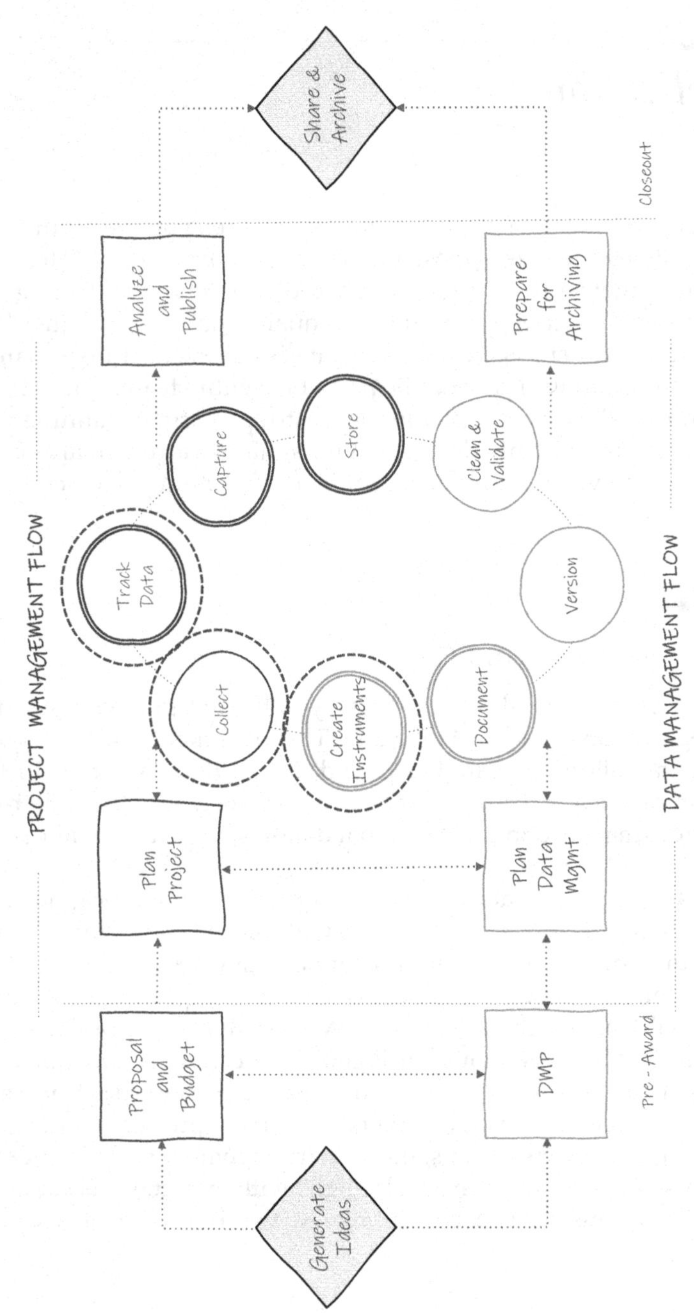

FIGURE 11.1
Data collection in the research project life cycle.

both project management and data management team members can implement that will improve the integrity of your data.

1. Quality assurance practices that happen before data is collected.
   - Best practices associated with designing and building your data collection instruments.
2. Quality control practices implemented during data collection.
   - Best practices associated with managing and reviewing data during collection.

11.2 Quality Assurance

Education researchers collect original data in many ways (see Figure 11.2). The focus of this chapter will be on data collected via forms (i.e., a document with spaces to respond to questions). Forms are widely used to collect data in education research (i.e., think questionnaires, assessments, observation forms, or a progress monitoring form), yet if poorly developed, they can produce some of the most problematic data issues.

The focus on forms is not to discount the importance of data collected through other means such as video or audio recording, where issues such as participant privacy and data security and integrity should absolutely also be considered. However, even with those types of data collection efforts, often teams are ultimately still coding that data using some sort of form (e.g., observation form), further supporting the need to build forms that collect quality data.

When collecting information using forms you can certainly do your best to fix data errors after data collection during a cleaning process. However, one of the most effective ways to ensure quality data is to correct it at the source.

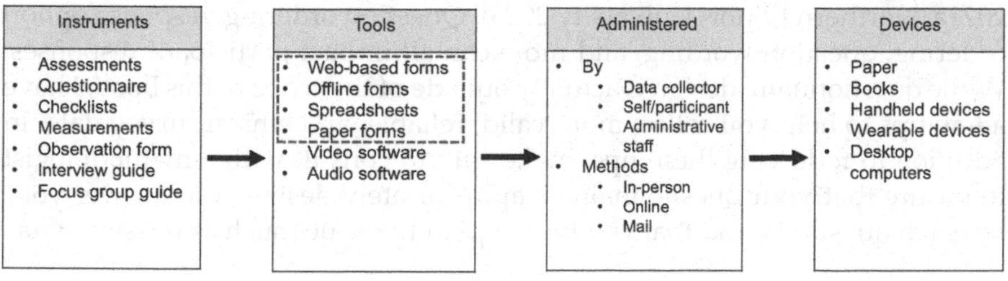

FIGURE 11.2
Common education research data collection methods.

This means designing items and building data collection tools in a way that produces valid, reliable, and more secure data. When creating your original data collection instruments, there are five ways to collect higher quality data.

1. Using good questionnaire design principles
2. Implementing a series of pilot tests
3. Choosing data collection tools that meet your needs
4. Building your instrument with the end in mind
5. Ensuring compliance

We will discuss each of these phases below.

> **NOTE**
>
> If you are collecting data using a standardized assessment, along with a provided tool (e.g., a computer-adaptive testing program, a testing booklet), most of the information in this section will not be applicable as it is best to adhere to all guidelines provided by the assessment company. You can skip to Section 11.2.5.

11.2.1 Questionnaire Design

In Chapter 8 we discussed the importance of documenting all instrument items in your data dictionary before creating your data collection instruments. As you develop items to add to each data dictionary for each original data source, it is vital to consider questionnaire design.

While some instruments (e.g., cognitive assessments) typically have standardized items, other instruments, such as surveys, are often not predefined, allowing researchers freedom in the design of the instrument which can lead to negative effects such as errors, bias, and potential harm (DIME Analytics 2021a; Northern Illinois University 2023). Question ordering, response option ordering, question wording, and more can all impact participant responses. While questionnaire design is actually outside of the scope of this book, I have a few tips to help you collect more valid, reliable, and ethical survey data. In addition to following these tips, make sure to consult with a methodologist to ensure that your questionnaire is appropriately designed to answer your research questions, and that you have a plan for issues such as missing data.

1. Follow technical documentation for existing scales.
    - If using an existing scale, as discussed in Section 8.4.1.1, make sure to follow any technical documentation associated with

  that scale (e.g., item wording, response options, item order). This ensures that you are using the scale as intended, and also improves opportunities to integrate data or replicate findings across other studies that have used the same scales.
- If you have used a measure before in a previous project, it is also beneficial to keep wording and response options consistent across projects for similar reasons.

2. Use existing standards if possible.
   - Organizations such as the National Institutes of Health and the National Center for Education Statistics have developed repositories (Common Data Elements[1] and Common Education Data Standards[2]) of standardized question wording paired with a set of allowable response options for commonly used data elements. Using standards when collecting variables, such as demographics, provides the following benefits (ICPSR 2022; Kush et al. 2020):
     - Reduces bias.
     - Allows for harmonization of data across your own research studies and also across the field.
       - This allows researchers to draw conclusions using larger samples or by comparing data over time.
       - It also reduces the costs of integrating datasets.
     - Improves interpretation of information.
3. Make sure questions are clearly worded and answer choices are clear and comprehensive.
   - Consider how the language might be interpreted. Is the question wording confusing? Can the response options be misinterpreted?
     - Rather than asking "What county are you from?" when looking for the participant's current location, be more specific and ask, "What county do you currently reside in?".
     - Rather than asking "Which parent are you?" and providing the response options "m" and "f" (where "m" and "f" could be interpreted as "male" or "female") clearly write out the response options and make sure they are comprehensive ("mother", "father", "legal guardian", and so forth).
     - Rather than asking "Do your children not have siblings?" which can be confusing, remove the negative and ask, "Do your children have siblings?" (T. Reynolds, Schatschneider, and Logan 2022).
   - Is the question leading/biased?
     - Are the response options ordered in a leading way?

- Is there more than one way to answer this question?
  - Are response categories mutually exclusive and exhaustive (ICPSR 2020)?

4. Consider data ethics in your questionnaire design (Frederick 2020; Gaddy and Scott 2020; Kaplowitz and Johnson 2020; Kopper and Parry 2023b; Mathematica 2023).
   - Consider the why of each item and tie your questions to outcomes.
     - Don't cause undue burden on participants by collecting more data just to have more data.
     - If collecting demographic information, provide an explanation of why that information is necessary and how it will be used in your research.
   - Review question wording.
     - Does it have potential to do harm to participants? Do the benefits outweigh the risks?
     - If sensitive questions are included, make sure to discuss how you will protect respondent's information.
   - Make questions inclusive of the population while also capturing the categories relevant for research.
     - For demographic information, allow participants to select more than one option.
     - If a demographic question is multiple choice, still include an "other" or "prefer to self-describe" option with an open-text field.
   - Consider including one general free-text field in your survey to allow participants to provide additional information that they feel was not captured elsewhere.
5. Limit the collection of personally identifiable information (PII).
   - Collecting identifiable information is a balancing act between protecting participant confidentiality and collecting the information necessary to implement a study. We often need to collect some identifying information either for the purposes of record linking or for purposes related to study outcomes (e.g., scoring an assessment based on participant's age).
   - As a general rule, you only want to collect PII that is absolutely necessary for your project, and no more (Gaddy and Scott 2020). As discussed in Chapter 4, PII can include both direct identifiers (e.g., name or email) as well as indirect identifiers (e.g., date of birth). Before sharing your data, all PII will need to be removed or altered to protect confidentiality (see Section 16.2.3.4).

Resources

Source	Resource
Jessica Kay Flake, Eiko Fried	Measurement Schmeasurement: Questionable Measurement Practices and How to Avoid Them[3]
Pew Research Center	Writing survey questions[4]
Sarah Kopper, Katie Parry	Survey design[5]
Stefanie Stantcheva	How to run surveys: A guide to creating your own identifying variation and revealing the invisible[6]
World Bank	Survey content-focused pilot checklist[7]

11.2.2 Pilot the Instrument

Gathering feedback on your instruments is an integral part to the quality assurance process. There are three phases to piloting an instrument (DIME Analytics 2021b) (see Figure 11.3).

1. Gathering internal feedback on items
   - As discussed in Section 8.4.1, once all items for each instrument have been added to a data dictionary, have your data management working group (DMWG) (see Chapter 6) review the data dictionary and provide feedback.
2. Piloting an instrument for content
   - Once your DMWG has approved the items to be collected, the second phase of piloting can begin. Create a printable draft of your instrument that can be shared with people in your study population and gather feedback. Consult with your IRB to determine if approval is required before piloting your instrument with your study population.
3. Piloting the instrument for data related issues
   - Once the instrument is created in your chosen data collection tool, share the instrument with your team for review. Here we are most interested in whether or not the data we are collecting are accurate, comprehensive, and usable. We will discuss this phase in greater detail in Section 11.2.4.

Last, as you move through the piloting phases, remember to make updates not only in your tool but also in your data dictionary and any other relevant documentation (e.g., data cleaning plan).

Pilot Phase	Phase 1 - Team feedback	Phase 2 - Content	Phase 3 - Data
When	Data documentation phase	Create instruments phase	Create instruments phase
Who tests	Team members	People from your study population	Team members
What to provide to testers	A data dictionary for your instrument	All instrument items in the planned order on paper	Fully built instrument in chosen tool (e.g., web-based platform, paper form)
Example items to include in a feedback checklist	- Are all items included? - Are we in agreement about how items are named? - Are the items worded correctly? - Are response options correct? - Are response options coded correctly?	- Are the items clearly worded? - Are any items sensitive? - Are answer choices comprehensive? - Is the item order clear? - Is the time to complete survey acceptable?	- Were there any barriers to accessing the instrument? - Are all questions accounted for? - Are the items worded correctly? - Are all response options visible for each categorical question? - Were you able to enter unallowable values, data types, or formats? - Is the branching logic working? - Are you allowed to skip items that you should not be able to?
Next steps	Update changes in data dictionary. Create a paper draft of your instrument. Move to Phase 2.	Update changes in data dictionary. Create full instrument in chosen tool before moving on to Phase 3.	If data is collected electronically, export sample data and review it. Update changes in data dictionary. Make edits to instrument based on feedback and findings from exported data.

FIGURE 11.3
Data collection instrument pilot phases.

11.2.3 Choose Quality Data Collection Tools

Once content piloting is completed, teams should be ready to begin building their instruments in their data collection tools (see Figure 11.2 for examples of tools). Research teams may be restricted in the tools they use to collect their data for a variety of reasons including limited resources, research design, the population being studied, sensitivity levels of data, or the chosen instrument (e.g., an existing assessment can only be collected using a provided tool). However, if you have the flexibility to choose how you collect your data, pick a tool that meets the various needs of your project while also providing data quality and security controls. Things to consider when choosing a data collection tool are:

1. Needs of your project
    - Is crowdsourcing required?
    - Is multi-site access required?
    - Who is entering the data (i.e., data collectors, participants)?
        - If participants are entering data, is the tool accessible for your population?
    - What are the technical requirements for the tool (i.e., will internet be available if you plan to use a web-based tool)?
    - Does the tool have customizable features that are necessary for your instrument (e.g., branching logic, automated email reminders, options to embed data, options to calculate scores in the tool)?
2. Compliance and security
    - Consider the classification level of each data source (see Chapter 4).
        - Which tools are approved by your institution to protect the sensitivity level of your data?
    - If collecting anonymous data, do you have the option to anonymize responses in the tool (e.g., remove IP Address and other identifying metadata collected by the tool)?
    - Does the tool include data backups?
3. Training needed
    - Is any additional team training needed to allow your team to use and/or build instruments in the tool?
4. Associated costs
    - Is there a cost associated with the tool? Do you have the budget for the tool?
    - Will there be additional costs down the line (e.g., collecting data on paper means someone will need to hand enter the data later)?

5. Data quality features
   - Does the tool allow you to set up data validation?
   - Does the tool have version control?
   - Does the tool have features to deal with fraud/bots?

While there are a variety of tool options, in a nutshell when it comes to data collected via forms, data collection tools can be categorized in one of two ways—electronic or paper. In addition to choosing tools based on the criteria mentioned in this section, there are some general benefits associated with each method that should also be considered, especially when the research team has control over how the data collection tool is built (Cohen, Manion, and Morrison 2007; Douglas, Ewell, and Brauer 2023; Gibson 2021; ICPSR 2020; Malow et al. 2021; Society of Critical Care Medicine 2018; Van Bochove, Alper, and Gu 2023) (see Table 11.1).

NOTE

If you choose to collect data in an electronic (also called digital) format, I highly recommend using a web-based tool that directly feeds into a shared database rather than through offline tools that store data on individual devices. Using a web-based tool, all data is stored remotely in the same database and can be easily downloaded or connected to at any time. No additional work is required.

However, when collecting data on various tablets in the field, if the forms are offline and cannot be later connected to a web-based form, then all data will be stored individually on each tablet. This not only may be less secure (e.g., a tablet becomes corrupted), it may also require additional data wrangling work including downloading data from each tablet to a secure storage location each day and then combining all files into a single dataset. If you use an electronic tool but your site does not have internet, consider using one of the many tools (e.g., Qualtrics, SurveyCTO) that allow you to collect data using their offline app and then upload that data back to the platform once you have an internet connection again.

Resources

Source	Resource
Michael Gibson, Wim Louw	Survey platform comparison[8]
Washington State University Libraries	Software for sensitive data[9]
Benjamin Douglas, et al.	Data quality in online human-subjects research comparison of tools[10]

TABLE 11.1
Comparison of Data Collection Tool Benefits

Electronic Data Collection Benefits	Paper Data Collection Benefits
Scalable (easier to reuse, edit, and maintain)	Intuitive to create and use (no training required)
Efficient (reduces both cost and effort associated with printing, collecting, and entering data)	Easy to do cognitive checks (eyeball for errors)
Prevents inconsistencies in data (e.g., using logic checks and data validation)	Easier to catch errors early on (in the field)
Reduces the chances of missing data (e.g., using response validation)	Can be more accessible for certain populations (e.g., young children)
Opportunity to reduce bias (e.g., through random question ordering)	
Potential to reach broader populations (e.g., crowdsourcing)	
Quicker turnaround of analysis-ready data and provides the opportunity to build real-time reporting pipelines (e.g., using APIs)	

11.2.4 Build with the End in Mind

As you create your data collection tool, you will want to build it with the end in mind. This means taking time to consider how the data you collect will be translated into a dataset (Beals and Schectman 2014; Lewis 2022b; UK Data Service 2023). Recall from Chapter 3, we ultimately need our data to be in a rectangular format, organized according to the basic data organization rules, in order to be analyzable. The process for building your tools with the end in mind is fairly different for electronic tools compared to paper forms, so we are going to talk about these two processes separately.

11.2.4.1 Electronic Data Collection

If you have ever created a data collection instrument and expected it to export data that looks like the image on the left of Figure 11.4, but instead you export data that looks like the image on the right, then you understand how important it is to spend time planning how data will be collected in your tool.

The first thing you will want to do before building your tool is bring out your data dictionary. This data dictionary will be your guide as you build your instrument. Some tools, such as REDCap, provide the option to upload your data dictionary which can then be used to automate the creation of data collection forms as opposed to building them from scratch (Patridge and Bardyn 2018). However, if you are building your instrument manually, adhering to the following guidelines will ensure you collect data that is easier

Data collected with planning

sch_name	tch_years	stress1	stress2
Silver Oak Elementary	2	1	3
Silver Oak Elementary	10	4	1
Sun Valley Middle	3	2	2
Sun Valley Middle	1	5	5

Data collected without planning

Q1	Q2	Q3	Q6
Silver Oak Elmenatry	two years	1	
Silver Oak	10	14	
Sun Valley	2 years high school, 1 year middle	2	2
Sun Valley Middle	1 yr 2 months	15	5

FIGURE 11.4
A comparison of data collected without planning and data collected with planning.

to interpret and more usable, and it will also reduce the amount of time you will need to spend on future data cleaning (Lewis 2022b).

1. Include all items from your data dictionary.
   - In addition to all original data collection items, this also includes any variables that will be derived during data collection either automatically in your tool or manually entered by data collectors (e.g., total correct, age calculation).
   - This does not, however, include any variables that you plan to derive or add to the data after data collection, during the data cleaning phase (e.g., treatment, standard scores).
2. Name all of your items the correct variable name from your data dictionary (UK Data Service 2023).
   - For example, instead of using the platform default name of `Q2`, rename the item to `tch_years`.
   - As mentioned in Section 9.4, it's also best to not concatenate a time component to your variable names if your project is longitudinal. Doing so makes it difficult to reuse your instrument for other time periods, creating additional work for you or your team.
3. Code all values as they are in your data dictionary.
   - For example, "strongly agree" = 1 | "agree" = 2 | "disagree" = 3 | "strongly disagree" = 4
   - Many times, tools assign a default value to your response options and these values may not align with what you've designated in your data dictionary.
   - As you edit your survey, continue to check that your coded values did not change due to reordering, removal, or addition of new response options.

4. Add data validation to reduce errors and missing data (UK Data Service 2023).
    - Content validation for open-text boxes.
        - Restrict entry to the variable type assigned in your data dictionary (e.g., numeric).
        - Restrict entry to the format assigned in your data dictionary (e.g., *YYYY-MM-DD*).
        - Restrict ranges based on allowable ranges in your data dictionary (e.g., *1–50*).
            - This could even include validating against previous responses (e.g., if "SchoolA" was selected in a previous question, grade level should be between *6 and 8*, if "SchoolB" was selected, grade level should be between *7 and 8*).
    - Response validation
        - Consider the use of forced-response and request-response options to reduce missing data.
            - Forced-response options do not allow participants to move forward without completing an item. Request-response options notify a respondent if they skip a question and ask if they still would like to move forward without responding.
            - Be aware that adding a forced-response option to sensitive questions has the potential to be harmful and produce bad data. If adding a forced-response option to a sensitive question, consider allowing those participants to opt-out in another way (e.g., "Prefer not to answer") (Kaplowitz and Johnson 2020; Kopper and Parry 2023b).
5. Choose an appropriate question type and format to display each item.
    - Become familiar with the various question types available in your tool (e.g., rank order, multiple choice, text box, slider scale).
    - Become familiar with the various formats (e.g., radio button, drop-down, checkbox).
    - For example, if your item is a rank order question (ranking three items), creating this question as a multi-line, free-text entry form may lead to duplicate entries (such as entering a rank of *1* more than once). However, using something like a rank order question type with a drag and drop format ensures that participants are not allowed to duplicate rankings.
6. If there is a finite number of response options for an item, and the number isn't too large (less than ~ 20), use controlled vocabularies

(i.e., a predefined list of values) rather than an open-text field (OpenAIRE_eu 2018; UK Data Service 2023).

- For example, list school name as a drop-down item rather than having participants enter a school name.
    - This prevents variation in text entry (e.g., "Sunvalley Middle", "sunvalley", "Snvally Middle"), which ultimately creates unnecessary data cleaning work and may even lead to unusable values.

7. If there is an infinite number of response options for an item or the number of options is large, use an open-text box.
    - If you can create a searchable field in your tool, allowing your participants to easily sift through all of the options, you absolutely should. Otherwise, use a text-box as opposed to having participants scroll through a large list of options.
    - Consider adding examples of possible response options to clarify what you are looking for.
    - Using open-ended text boxes does not mean you cannot regroup this information into categories later during a cleaning process. It is just more time-consuming and requires interpretation and decision-making on the part of the data cleaner.
8. Only ask for one piece of information per question.
    - Separating information prevents confusion in case a participant or data collector swaps the order of information.
    - For example, rather than asking "Please list the number of students in your algebra class and geometry class", split those into two separate questions so those questions download as two separate items in your dataset: "Please list the number in algebra class"; "please list the number in geometry class".
    - This also includes more simple examples such as splitting `first_name` and `last_name` into two separate fields.
9. To protect participant privacy and ensure the integrity of data, consider adding a line to the introduction of your web-based instrument, instructing participants to close their browser upon completion so that others may not access their responses.
10. Last, if possible, export the instrument to a human-readable document to perform final checks.
    - Are all questions accounted for?
    - Are all response options accounted for and coded as they should be?
    - Is skip logic shown as expected?

Once your tool is created, the last step is to pilot for data issues (see Figure 11.3). Collect sample responses from team members. Create a feedback checklist for them to complete as they review the instrument (Gibson and Louw 2020). Assign different reviewers to enter the survey using varying criteria (e.g., different schools, different grade levels, other branching options). Let team members know that they should actively try to break things—try to enter nonsensical values, try to skip items, try to enter duplicate entries (Kopper and Parry 2023a). If there are problems with the tool, now is the time to find out.

After sample responses are collected from team members, export the sample data using your chosen data capture process (see Chapter 12). Comparing the export to your data dictionary, review the data for the following:

1. Are there any unexpected or missing variables?
2. Are there any unexpected variable names?
3. Are there unexpected values for variables?
4. Are there missing values where you expect data?
5. Are there unexpected variable formats?
6. Is data exporting in an analyzable, rectangular format (see Chapter 3)?

If any issues are found either through team feedback or while reviewing the exported sample data, take time to update the tool as needed before starting data collection. This is also the time to update your documentation. As you review your exported file, update your data dictionary to reflect any unexpected variables that are included (e.g., metadata), any unexpected formatting, as well as any newly discovered recoding or calculations that will be required during the data cleaning process. As an example, if upon downloading your sample data, you learn that a "select all" question exports differently than you expected, now is the time to add this information, along with any necessary future transformations, to your data dictionary. This is also a great time to update your data cleaning plan (see Section 8.3.3) with any new transformations that will be required.

11.2.4.2 Paper Data Collection

There are many situations where collecting data electronically may not be feasible or the best option for your project. While it is definitely trickier to design a paper tool in a way that prevents bad data, there are still steps you can take to improve data quality.

1. Use your data dictionary as a guide as you create your paper form.
    - Make sure all questions are included and all response options are accurately added to the form.

2. Have clear instructions for how to complete the paper form (Kopper and Parry 2023b).
    - Make sure to not only have overall instructions at the top of the form but also have explicit instructions for how each question should be completed.
        - Where to write answers (e.g., not in the margin)
        - How answers should be recorded (e.g., *YYYY-MM-DD*, 3-digit number)
        - How many answers should be recorded (e.g., circle only one answer, check all applicable boxes)
        - How to navigate branching logic (e.g., include visual arrows)
3. Only ask for one piece of information per question to reduce confusion in interpretation.

Once your tool is created, you will want to pilot the instrument with your team for data issues (see Figure 11.4). Using the feedback collected, edit your tool as needed before sending it out into the field.

Last, unless paper data are collected using a machine-readable form, they will need to be manually entered into an electronic format during the data capture phase. While we will talk about data entry specifically in Section 12.2, this point in instrument creation is a great time to create an annotated instrument (Neild, Robinson, and Agufa 2022). This includes taking a copy of your instrument and writing the associated codes alongside each item (i.e., variable name and value codes). This annotated instrument can be useful during the data entry process and serve as a linking key between your instrument and your data dictionary. See Figure 11.5 for an example of an annotated instrument from wave 1 of the Florida State Twin Registry project shared in the LDbase repository (Hart, Schatschneider, and Taylor 2021).

11.2.4.3 Identifiers

When building data collection tools, no matter if they are paper or electronic, it is vitally important to make sure you are collecting unique identifiers (Kopper and Parry 2023b). Whether you have participants enter a unique identifier into a form or you link study ID (see Section 10.4) to each form in some other way, it's important to not accidentally collect anonymous data. Without unique identifiers in your data, you will be unable to link data across time and forms. If possible, you want to avoid collecting names as unique identifiers for the following reasons (McKenzie 2010):

- To protect confidentiality, we want to use names as little as possible on forms.

Ent. 1: ___ Ent. 2: ___ ID: ________

Home Environment Measure

The first section of this questionnaire focuses mostly on demographic characteristics of the twins' family.

1. [hem1] a.) The person completing this questionnaire is the twins' (check one):

_1_ Biological mother
_2_ Biological father
_3_ Step mother
_4_ Step father
_5_ Other relative (e.g., grandmother, aunt, etc.)
_6_ Adoptive or foster parent
_7_ Other (please explain: ______ hem1t ______)

[hem31] b.) The other adult caregiver in the home is:

_1_ Biological mother
_2_ Biological father
_3_ Step mother
_4_ Step father
_5_ Other relative (e.g., grandmother, aunt, etc.)
_6_ Adoptive or foster parent
_7_ Other (please explain: ______ hem31t ______)
_8_ N/A (there is no other adult caregiver in the home)

FIGURE 11.5
Annotated parent survey from wave 1 of the Florida State Twin Registry project (2018).

    - If they are used on forms, we want to remove them as soon as possible.
- Names are not unique.
    - If you do collect names, you'll want to ask for additional identifying information that when combined, make a participant unique (e.g., student name and email).
- Names change (e.g., someone gets married/divorced).
- There is too much room for error.
    - If names are hand entered, there are endless issues with case sensitivity, spelling errors, special characters, spacing, and so forth.

All of these issues make it very difficult to link data. If you do decide to collect names, remember that you will need to remove names during data processing and replace them with your unique study identifiers (see Section 14.3 for more information about this process).

Rather than having to de-identify your data through this cleaning process, another option is to collect a different type of unique identifier, or pre-link unique study identifiers and names in your instrument, removing many of these issues (DIME Analytics 2021a; Gibson and Louw 2020). We will discuss these methods separately for electronic data and paper data.

NOTE

If your study is designed to collect anonymous data, then you will not assign study identifiers and no participant identifying information should be collected in your instruments (e.g., name, email, date of birth). You will also want to make sure that if your tool collects identifying metadata, such as IP Address or worker IDs in the case of crowdsourcing tools (e.g., MTurk), this information will not be included in your downloaded data.

Remember that if you collect anonymous data, you will not be able to link data across measures or across time. However, if your study randomizes participants by an entity (e.g., school or district), you will need to collect identifying information from that entity in order to cluster on that information (e.g., school name).

11.2.4.3.1 Electronic Data

When collecting electronic data, there are many ways you might consider collecting unique identifiers other than names. A few possible options are provided below. The method you choose will depend on your data collection design, your participant population, your tool capabilities, and your team expertise.

1. Create unique links for participants.
    - Many tools will allow you to preload a contact list of participants (from your participant tracking database) that includes both their names and study IDs. Using this list, the tool can create unique links for each participant. This is the most error-proof way to ensure study IDs are entered correctly.
    - When you export your data, the correct ID is already linked to each participant and you can choose to not export identifying information (e.g., names, emails) in the data.

- If using this method, make sure to build a data check into the system. For example, when a participant opens their unique link, verify their identity by asking, "Are you {first name}?" or "Are your initials {initials}?" In order to protect participant identities, do not share full names.
    - If they say yes, they move forward. If they say no, the system redirects them to someone to contact. This ensures that participants are not completing someone else's survey and IDs are connected to the correct participant.

2. Provide one link to all participants and separately, in an email, in person, or by mail, provide participants with their study ID to enter into the system.
    - This might be a preferred method if you are collecting data in a computer lab or on tablets at a school site, or if your tool does not have the option to create unique links.
    - This can possibly introduce error if a participant enters their study ID incorrectly.
        - Similar to the first option, after a participant enters their ID, verify their identity.
    - Note that participants are only becoming aware of their own study identifier, not the identifiers associated with other participants. However, if your team, or your Institutional Review Board (IRB), is uncomfortable with participants knowing their study IDs you can also consider using a "double ID" which is yet another set of unchanging unique identifiers that you use for the sole purpose of data collection. Those identifiers will need to be tracked in your participant tracking database and will need to be replaced with study IDs in the clean data.
3. If you have not previously assigned study identifiers (e.g., your consent and assent process is a part of your instrument), you can have participants enter their identifying information (e.g., name) and then have the tool assign a unique identifier to the participants.
    - Using this method, you can potentially download two separate files.
        - One with just the instrument data and assigned study ID, with name removed
        - One with just identifying information and assigned study ID (this information will be added to your participant tracking database)

11.2.4.3.2 Paper Data

If you take paper forms into the field consider doing the following to connect your data to a participant (O'Toole et al. 2018; Reynolds, Schatschneider, and Logan 2022).

- Write the study ID, and any other relevant identifiers (e.g., school ID and teacher ID), on each page of your data collection form and then attach a cover sheet with participant name and other relevant information (see Figure 11.6). When you return to the office, you can remove the cover sheet and be left with only the ID on the form.
  - It is this ID only that you will enter into your data entry form during the data capture process, no name.
  - Removing the cover sheet ensures that your data entry team only sees the study ID when they enter data, increasing privacy by minimizing the number of people who see participant names.
  - However, before removing the cover sheet, double-check study identifiers against your participant database to make sure the information is correct.
  - Make a plan for the cover sheets (either shred them if they are no longer needed, or store them securely in a locked file cabinet and shred them at a later point).

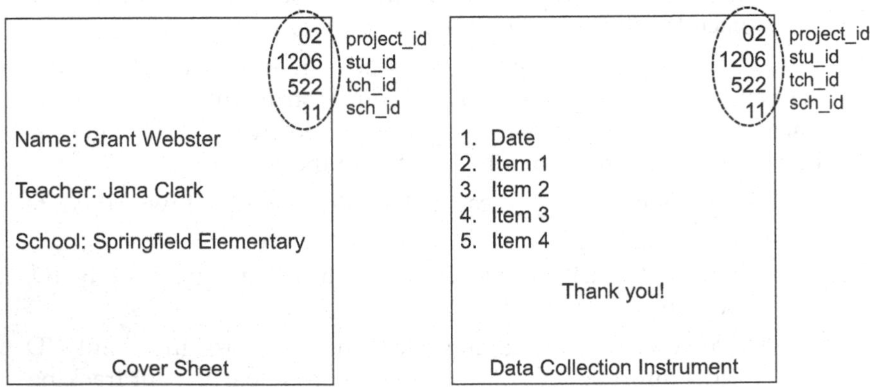

FIGURE 11.6
Example cover sheet for a paper data collection instrument.

11.2.5 Ensure Compliance

If you are collecting human subjects data and your study is considered research (see the Glossary for definitions of these terms), it is important to consult with your applicable IRB about their specific requirements before moving forward with any data collection efforts. As discussed in Section 4.3.2, an IRB is a committee that assesses the ethics and safety of research studies involving human subjects. If an IRB application is required for your project, the review process can take several weeks or more, and it is common for the IRB to request revisions to submission materials. Make sure to review your timeline and give yourself plenty of time to work through this process before you need to begin recruitment and data collection.

Informed consent agreements, and assents for participants under the age of 18, are commonly required by IRBs for research studies that collect human subjects data. As mentioned in Section 4.3.4, these agreements ensure that participants fully understand what is being asked of them and voluntarily agree to participate in your study. There are several categories of information that may be required for you to include in your consent form (e.g., description of study, types of data being collected, risks and benefits to participant, how participant privacy will be maintained) (The Turing Way Community 2022). Make sure to consult with your applicable IRB about their specific requirements. With an increase in federal data sharing requirements, it is very important at this time to also consider how you want to gain consent for public data sharing (Levenstein and Lyle 2018). There are two risks posed by data sharing and it is important to address both in your consent—loss of privacy and using data for purposes participants did not agree to. Meyer (2018) provides some general best practices to consider when adding language about public data sharing to a consent form.

- Don't promise to destroy your data (unless your funder/IRB explicitly requires it).
  - Do incorporate data retention and sharing plans including letting participants know who will have access to their data.
- Don't promise to not share data.
  - Do get consent to retain and share data (consider adding the specific repository you plan to share your data in).
  - Consider offering tiered levels of consent for participants who may not want all of their data publicly shared but will allow some.
- Don't promise that research analyses of the collected data will be limited to certain topics.
  - Do say that data may be used for future research and share general purposes (e.g., replication, new analyses).
- Do review the ways you plan to de-identify data and be thoughtful when considering risks of re-identification (e.g., small sample size for subgroups).

There are essentially three different ways you can go about obtaining consent for data sharing (Gilmore, Kennedy, and Adolph 2018).

1. Include a line about public data sharing in your consent to participate in research.
    - With this method, a participant who consents is agreeing to both participate in the research study and have their data shared publicly.
2. Have participants consent to data sharing at the same time that you provide the research study consent, but provide a separate consent form for the purposes of public data sharing.
3. Have participants consent to data sharing on a separate consent form, at a later time, after research activities are completed.
    - Obtaining consent this way ensures the participants are fully aware of the data collected from them and can make an informed decision about the future of that data.

Consult with your IRB to determine the preferred method for obtaining consent for public data sharing. If you use method 2 or 3, it is very important that you not only track your participant study consent status in your tracking database (as discussed in Chapter 10), but that you also add a field to track the consent status for data sharing so that you only publicly share data for those that have given you permission to do so. You will also want to consider who is included in your final analysis sample. If you include all consented participants in your analysis, your publicly available dataset will not match your analysis sample if some people did not consent to data sharing. You may need to consider options such as using a controlled access repository to share the full sample for purposes of replication. We will discuss different methods of data sharing in Chapter 16.

11.2.5.1 Building Consent Forms

If you are required to collect consent for your study, quality assurance should be considered with these forms as well. Keeping consent and assent forms secure should be a top priority for your team due to the identifiable information on the form. It is also important to be able to clearly identify the consenting individual on the form. Whether you collect consent on paper or electronically, make sure you have a clear quality assurance plan.

1. Use institution and IRB approved tools to collect consent.
2. If collecting paper consent, make sure that you will be able to clearly identify the consenting individual (e.g., participant printed name or

signature alone may not be sufficient due to duplicate names, nicknames used, or illegible handwriting). One option is to pre-print names and other relevant information on forms or have school staff write participant names on forms before handing them out.

Templates and Resources	
Source	**Resource**
Anja Sautmann	Annotated informed consent checklist[11]
Holly Lane, Wilhemina van Dijk	Example parent consent[12]
ICPSR	Recommended Informed Consent Language for Data Sharing[13]
Jeffrey Shero, et al.	Informed consent and waiver of consent cheat sheet[14]
Jeffrey Shero, Sara Hart	Informed consent template with a focus on data sharing[15]
Melissa Kline Struhl	Lookit consent form template 5[16]
University of Virginia	A collection of consent and assent templates[17]

11.3 Quality Control

In addition to implementing quality assurance measures before data collection, it is equally important to implement several quality control measures while data collection is underway. Those measures include:

1. Field data management
2. Ongoing data checks
3. Tracking data collection daily
4. Collecting data consistently

We will discuss each of these measures in this section.

11.3.1 Field Data Management

If your data collection efforts include field data collection (e.g., data collectors administering assessments in a school), there are several steps your team can implement that will keep your data more secure in the field, help a project coordinator keep better track of what happens in the field, and will lead to more accurate and usable data. Some best practices for field data collection include the following (DIME Analytics 2021a):

- Keep your data secure in the field.
    - Make sure all paper forms are kept in a folder (or even a lock box) with you at all times, and that they are promptly returned to the office (e.g., not left in a car, not left at someone's home).
        - This is especially important when considering paper consent forms. Not only do these forms contain identifiable participant information, but if this form is misplaced, you no longer have consent to collect a participant's data.
    - Ensure all electronic data collection devices (e.g., phones, tablets) are password protected and never left open and unattended. Keep all identifiable information encrypted on your field devices (i.e., data is encoded so that only those with a password can decipher it). You may also consider remote wiping capabilities on portable devices in the case of loss or theft (O'Toole et al. 2018).
- Create tracking sheets to use in the field.
    - These sheets should include the names and/or identifiers of every participant who data collectors will be collecting data from.
    - Next to each participant, include any other relevant information to track such as
        - Was the data collected (i.e., a check box)
        - Who collected the data (i.e., data collector initials or ID)
        - Date the data was collected
        - As well as a notes section to describe any potential issues with the data (e.g., "Student had to leave the classroom halfway through the assessment" or "Student refused to continue assessment")
    - This tracking sheet allows the project coordinator to keep track of what is occurring in the field so that information can be accurately recorded in the participant tracking database and forms can be sent back out for completion as needed.
- Check physical data in the field.
    - Immediately upon completing a form, have data collectors do spot checks. If any problems are found, follow up with the participant for correction if possible.
        - Check for missing data
        - Check for duplicate answers given (i.e., two answers circled for a question when there should only be one)
        - Check for answers provided outside of the assigned area (e.g., answers written in the margins)

FIGURE 11.7
A series of spot checks that occur with paper data.

    - Check calculations and scoring (e.g., basals, ceilings, raw scores)

- Assign a field supervisor. This person is assigned to:
    - Do another round of data checks in the field once the data collector returns physical forms to the on-site central location (e.g., if data collectors have set up in the teacher's lounge).
    - Ensure that all data and equipment is accounted for and returned to the office.
    - Be available for trouble shooting as needed.
- Do another round of physical data spot checking as soon as the data is returned to the office (see Figure 11.7).
    - The project coordinator may do this round of checking as they are tracking information in the participant database.
    - If any issues are found, note that in the tracking database and send the form back out to the field for correction.
    - If paper forms are mailed back to you from participants, rather than returned from field data collectors, it is still important to do in-office spot checks. If at all possible, reach out to those participants for any corrections.
- When a wave of data collection wraps up, collect feedback from data collectors to improve future data collection efforts.
    - What went well? What didn't?

Templates	
Source	**Resource**
Crystal Lewis	Field tracking sheet template[18]

11.3.2 Ongoing Data Checks

If you collect data via a web-based form, you will want to perform frequent data quality checks, similar to the checks you performed during the content and data piloting phase. You will want to check for both programming errors (e.g., skip logic programmed incorrectly) as well as response quality errors (e.g., bots, survey comprehension) (DIME Analytics 2021a; Gibson 2021). Consider the following:

- Checks for comprehension
    - Are any questions being misinterpreted? If your form contains a free-text field, spot-check to see if responses make sense.
- Checks for missing data
    - Are items being skipped that should not be skipped?
    - Are participants/data collectors not finishing forms?
- Checks for ranges and formats
    - Are values in unexpected formats or falling outside of expected ranges?
- Checks for duplicate forms
    - Are there duplicate entries for participants?
- Is skip logic working as expected?
    - Are people being directed to the correct location based on their responses to items?

Some of these checks can be performed programmatically (i.e., you can write a script in a program such as R to pull and check for things like out of range values on a recurring schedule). Other checks may require manual effort (e.g., such as downloading your data on a recurring schedule and reviewing open-ended questions for nonsensical responses). If errors are found, consider revising your instrument to prevent future errors if this is possible without jeopardizing the consistency of your data.

11.3.3 Tracking Data Collection

Throughout data collection your team should be tracking the completion of forms (e.g., consents, paperwork, data collection forms) in your participant database (see Chapter 10). This includes paper forms, electronic forms stored on devices, as well as web-based data coming in. Your team may designate one person to track data (e.g., the project coordinator), or they may designate multiple. If you are working across multiple sites, with multiple teams, you will most likely have one or more people at each site tracking data as it comes in.

Some tracking best practices include:

1. Track daily during data collection.
    - Do not wait until the end of data collection to track what data was collected for each participant.

    - Tracking daily helps ensure that you don't miss the opportunity to collect data that you *thought* you had but never actually collected.
2. Only track data that you physically have (paper or electronic).
    - Never track data as "complete" that someone just tells you they collected.
        - You can always mark this information in a "notes" field but do not track it as "complete" until you have the physical data.
3. Only track complete data as "complete".
    - Review all data before marking it as "complete".
    - For paper files, make tracking the step immediately after your spot-checking process. If a form is only partially completed and you plan to send it back out to the field for completion, mark this in the "notes" but do not mark it as "completed". If you have a "partially completed" option, you can mark this option.
    - For electronic files, make tracking the step immediately following your data quality checks. If a form is only partially completed, mark this appropriately in your tracking database.

As mentioned in Chapter 10, if you use a tool such as REDCap for both collection and tracking, you may not need to do any manual tracking. Completion fields may automatically update as data is collected. You'll need to determine what is needed based on your specific scenario.

11.3.4 Collecting Data Consistently

It is important to collect data consistently for the entire project to ensure interoperability and harmonization of information. Keep the following consistent across repeated collections of the same form (e.g., Spanish and English version of a form, Site A and Site B form, wave 1 and wave 2 of collecting a form):

- Variable names
    - Use the same names for the same items.
    - Remember that it's best to not add a time component to your variable names at this time (see Section 9.4.1 for more information).
- Variable types and formats
    - For example, if `gender` is collected as a numeric variable on one form, keep it as a numeric variable for all forms.
    - Or, if `svy_date` was collected as *YYYY-MM-DD* on one form, keep it the same format across forms.

- Value codes
  - Make sure response options are consistently coded using the same values (e.g., "no" = 0 | "yes" = 1) (see Section 9.5 for more information).
- Question type and format
  - For example, if a slider question was used for "Percent of time on homework", continue to ask that question using a slider question.

Failing to collect your data consistently has many consequences.

1. It can make it difficult or impossible to compare outcomes.
2. It makes your work less reproducible.
3. It reduces your ability to combine data (i.e., you cannot append dissimilar variables).
4. It can lead to errors in interpretation.

Last, collecting data consistently also means measuring things in the same way over time or across forms so that you don't bias your results. The slightest change in item wording or response options can result in dramatic changes to outcomes (ICPSR 2022; Pew Research Center 2023). If an item absolutely needs to be edited after the study begins (e.g., a response option was left off), version the edited variable (e.g., `stress1` and `stress1_v2`) and note when the variable was edited in your data dictionary. Similarly, if items are added to an instrument at a later point in your study, make sure to follow quality assurance procedures including adding these items to your data dictionary and noting when the items were added, before adding them to your instrument.

11.4 Bot Detection

All of the web-based data collection efforts in this chapter assume you are making a private link that you are sharing with a targeted list (e.g., students in a classroom, teachers in a school). However, there may be times when you need to publicly recruit and collect data for your study, and this opens your instrument up for a plethora of data quality issues. Bots, fraudulent data, and incoherent or synthetic responses are all issues that can plague your online data collection efforts, particularly with crowdsourcing platforms (Douglas, Ewell, and Brauer 2023; Veselovsky, Ribeiro, and West 2023;

Webb and Tangney 2022). If possible, avoid using public survey links. One possible workaround is to first create a public link with a screener. Then after participants are verified through the screener, send a private, unique link to the instrument.

If a workaround is not possible and you need to use a public link, some quality assurance suggestions that can help you both secure your instrument and detect fraud include the following (Arndt et al. 2022; Simone 2019; Teitcher et al. 2015):

- Not posting the link on social media
- Using CAPTCHA verification, or a CAPTCHA alternative, to distinguish human from machine
- Using tools that allow you to block suspicious geolocations
- Not automating payment upon survey completion
- Including open-ended questions
- Building attention/logic checks into the survey
- Asking some of the same questions twice (once early on and again at the end)

Last, implement quality control checks for bots or fraudulent responses before analyzing data and before providing payments to participants. The following types of things are worth looking into further:

- Forms being completed in a very short period of time
- Forms being collected from suspicious geolocations
- Duplicated or nonsensical responses to open-ended questions
- Nonsensical responses to attention or logic checking questions
- Inconsistent responses across repeated questions

11.5 Review

Recall from Chapter 6, we discussed designing and visualizing a data collection workflow during your planning phase. As we've learned from this chapter, errors can happen at any point in the workflow, so it is important to consider the entire data collection process holistically and integrate both quality assurance and quality control procedures throughout. Figure 11.8 helps us to see when these practices fit into the different phases our workflow.

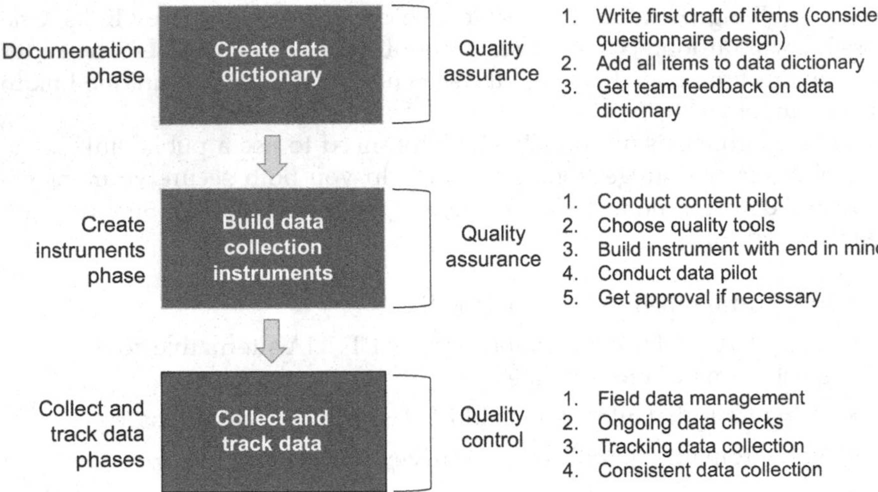

FIGURE 11.8
Integrating quality assurance and control into a data collection workflow.

Resources

Source	Resource
DIME Wiki	Questionnaire design timeline[19]
Sarah Kopper, Katie Parry	Five key steps in the process of survey design[20]

Once your workflow is developed and quality assurance and control practices are integrated, consider how you will ensure that your team implements these practices with fidelity. Document the specifics of your plan in an SOP (see Section 8.2.7), including assigning roles and responsibilities for each task in the process. Last, train your team on how to implement the data collection SOP, and implement refresher trainings as needed.

Notes

1 https://cde.nlm.nih.gov/home
2 https://ceds.ed.gov/
3 https://doi.org/10.1177/2515245920952393
4 https://www.pewresearch.org/our-methods/u-s-surveys/writing-survey-questions/
5 https://www.povertyactionlab.org/resource/survey-design

6 https://www.nber.org/system/files/working_papers/w30527/w30527.pdf
7 https://dimewiki.worldbank.org/Checklist:_Content-focused_Pilot
8 https://www.povertyactionlab.org/resource/survey-programming
9 https://libguides.libraries.wsu.edu/rdmlibguide/ethics
10 https://journals.plos.org/plosone/article?id=10.1371/journal.pone.0279720
11 https://www.povertyactionlab.org/sites/default/files/research-resources/rr_irb_annotated-informed-consent-checklist_0.pdf
12 https://www.ldbase.org/system/files/documents/2021-04/HS-Parent Consent.txt
13 https://www.icpsr.umich.edu/web/pages/datamanagement/confidentiality/conf-language.html
14 https://osf.io/3czbx
15 https://figshare.com/articles/preprint/Informed_Consent_Template/13218773
16 https://github.com/lookit/research-resources/blob/master/Legal/Lookit%20consent%20form%20template%205.md
17 https://research.virginia.edu/irb-sbs/consent-templates
18 https://osf.io/hd2w3
19 https://dimewiki.worldbank.org/Questionnaire_Design
20 https://www.povertyactionlab.org/resource/survey-design

12

Data Capture

After the data collection period is complete, the next phase in the cycle is to capture the data (see Figure 12.1). Here we are extracting, creating, or acquiring a flat file, consisting of data previously collected electronically or on paper, that we can then save in our designated storage location. In quantitative research we typically want to capture data in an electronic, rectangular format that can be easily analyzed or shared (see Chapter 3). In this chapter we will review common ways to capture data based on three data collection methods (see Figure 12.2). Similar to data collection, it is possible for data errors to occur during this phase. In reviewing data capture methods, we will also cover how data quality can be managed during this phase.

12.1 Electronic Data Capture

As discussed in Chapter 11, electronic data can be collected using a variety of software (either web-based or offline). Since electronic forms typically funnel data into a spreadsheet or database, it makes the process of data capture much easier compared to paper data. However, there is still much to consider.

1. How will the data be captured?
    - The most common way to capture web-based forms is to download it from your platform.
        - You may also be able to capture data via an API (application programming interface). If you regularly need to capture your data for quality control or other purposes, using an API can be a great way to remove the burden of manually logging into a program and going through the point and click process of downloading a file. Instead, you can write a script, in a program such as R, to extract the data. Once the script is created, you can run it as often as you want. However, this is only an option if your tool has an API available (e.g., Qualtrics).

DOI: 10.1201/9781032622835-12

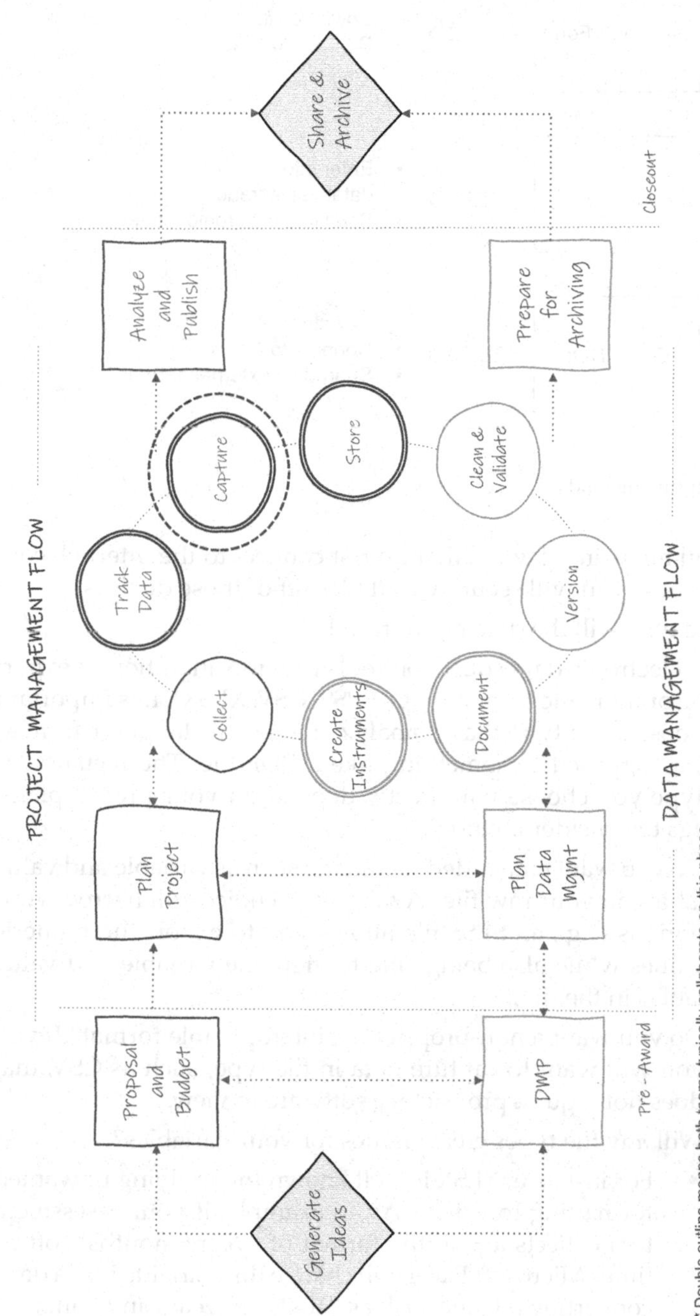

FIGURE 12.1
Data capture in the research project life cycle.

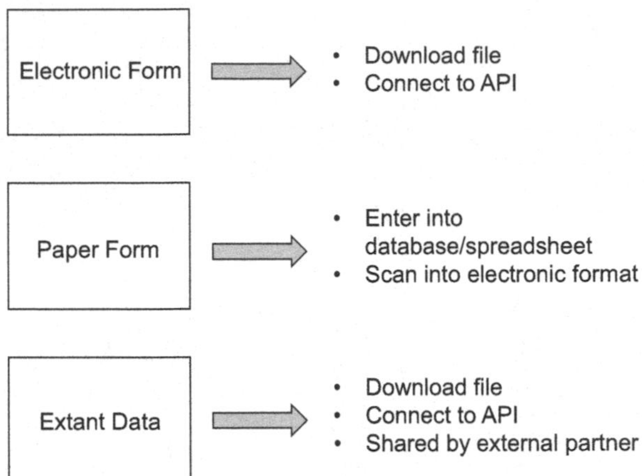

FIGURE 12.2
Common data capture methods.

- If you are using devices that do not connect to the internet, consider how you will securely pull files off of those devices.

2. What file type will the data be captured in?
   - Most electronic data collection tools provide an option to export to one or more file formats (e.g., SPSS, CSV, XLSX). It is important to choose a file type that is analyzable (i.e., rectangular formatted), as opposed to something like a PDF file. The rectangular file type you choose will mostly depend on your project plans. Things to consider include:
     - Do you want embedded metadata, such as variable and value labels, in your raw file? Again, your choice will narrow your options (e.g., an SPSS file allows you to export the numeric values while also being able to view the variable and value labels in the file).
     - Do you want a non-proprietary, interoperable format? If yes, you will want to capture data in file type, such as CSV, that does not require proprietary software to view.
     - Will any file types create issues for your variables?
       - For instance, XLSX is well known for applying unwanted formatting to values. As an example, if your assessment tool collects age in the format of "years-months", oftentimes Microsoft Excel will change this variable into a date, converting a value such as "10-2" (10 years and 2 months old) to "2-Oct". A more suitable file type in this situation may be a CSV or TXT file, which do not apply formatting.

- Is there a file structure that you don't want to work with?
    - As an example, the structure of an SPSS file may look different compared to an XLSX file depending on the tool. In a tool like Qualtrics, an XLSX or CSV file may export with multiple header rows whereas an SPSS file does not.

3. What additional formatting options need to be considered?
    - First and foremost, take time to understand the default options of your tool. From there, your tool may provide alternatives to the default. The available options and how they are worded will vary depending on what tool you use. However, examples of options you may encounter include:
        - Options to export the text values for categorical fields or to export the numeric values associated with each category. Your options may be limited by your chosen file type (e.g., text values may be available in a CSV but not in an SPSS file).
        - Options for formatting "select all that apply" (or multiple response) questions. Typically tools provide two options (see Figure 12.3).
            1. Export all selections in one variable, where each selection is separated by a comma. As you can imagine, this can get messy if you export text responses that contain commas.
            2. Split each option into its own column.
        - Options for recoding seen but unanswered questions.
            - Some tools may allow you to recode these blanks to a value such as *-99*. Selecting this option enables you to better determine if values are missing because they were not seen or if they were seen and purposefully skipped. This can be especially helpful for multiple response questions where cell values are blank when an option is not selected. When no options are selected, it can be difficult to determine if the entire question was skipped or if options were purposefully not selected, unless seen but unanswered cells are recoded.
        - Options to not export variables tagged as identifiers.
            - If your tool allows you to tag certain fields as identifiable (e.g., name, email), you may have the option to exclude those variables in your export.
4. Where will the file be stored?
    - This decision should be based on guidelines you've laid out in your data storage documentation (see Chapter 13).

Export comma separated responses

tch_id	cert
235	Elementary
236	Middle Math,Secondary Math
237	
238	PreK, Primary,Elementary

Export responses as individual variables

tch_id	cert_1	cert_2	cert_3	cert_4
235		Elementary		
236			Middle Math	Secondary Math
237				
238	PreK, Primary	Elementary		

FIGURE 12.3
Two different ways you may export a "select all that apply" question.

5. How will files be named?
   - While your tool may provide a name for your file, it may need to be renamed something more meaningful based on your style guide rules (see Section 9.3). Most importantly, name your files consistently across data sources and waves.
6. What documentation needs to accompany the data capture?
   - A README can be very beneficial to store alongside the file if there is anything in the file that a future person managing the data should be aware of (see Section 8.3.1).
   - A changelog can also be very beneficial (see Section 8.3.2) to store with this file. It is common to have to redownload a raw data file due to errors found or new participants added. A changelog can help the team both identify the most recent version of a raw data file and understand the differences between files.
7. What checks need to happen before this data is handed off?
   - It is important for the person responsible for data capture to do a basic review of the file before handing this data off for the next step.
     - Does the format of the file look as expected? Does it have data in it? Are all the variables there as expected?
     - Are all participants in the data? This is an excellent time to compare the number of unique participants in the file to the number of participants with data marked as complete in your participant tracking database. If these numbers do not match, the person in charge of data capture should begin reconciling errors before handing off this data.
       - Was a participant accidentally dropped from the file? Is someone incorrectly marked as complete in the tracking database? Are there duplicate entries in the file?
         - If there are errors that can be corrected (e.g., someone incorrectly tracked a data point, a participant was left off in the capture process), those corrections should be made now. If there are corrections that involve manipulating the raw data (e.g., reconciling duplicate IDs in the data, an ID incorrectly entered in the data), those corrections should not be made at this time. Instead, those should be added to a README file to be corrected in the data cleaning phase.
8. Who will capture the data?
   - It doesn't necessarily matter who takes on this responsibility. What matters most is that the person has the expertise to capture

the data and that this responsibility is documented. If the person capturing the data is not the person who oversees data collection, it is important to still assign the data collection supervisor the responsibility of documenting any relevant information in a README.

NOTE

It is important to never make changes directly to the raw data files. This also includes not making changes directly to the data in your data collection tool. If you see errors in the raw data file that can't be fixed by simply re-downloading the data, make notes in a README for future correction as noted earlier in this section. Those corrections can be made in the data cleaning process. Doing this ensures that your raw data always reflects the true values that the participant reported and allows you to track data lineage. The one exception to this rule is if you accidentally collect data on a non-consented participant. In this case, it may be best to delete data for this participant directly in your data collection tool so that no record is kept.

12.1.1 Documenting Electronic Data Capture

All of these decisions should be made and documented during the time you are developing data collection tools. Making these decisions early allows you to also implement them during the pilot testing and data checking processes. For instance, if you plan to capture your data by exporting a CSV file from your data collection platform with a variety of options selected, you will want to use this same method during your data piloting and data checking process. This allows you to know exactly what your data will look like once data collection is complete and make adjustments as needed.

As discussed in Chapter 6, your data capture process should be added to your workflow diagram and then detailed in an SOP (see Section 8.2.7). All of the decisions in this process should exist in the relevant SOP. This ensures that workflows are standardized and reproducible. As we've learned in this section, one deviation from the SOP has the potential to produce a very different data product (e.g., the format of a CSV file compared to an SPSS file can vary). Not only can this produce errors, but it can also undermine the reproducibility of a data cleaning pipeline. Imagine a scenario where a data cleaning syntax is written to import a CSV file with an expected format, and that format changes. The pipeline is no longer reproducible. Last, documenting a timeline for when this data capture process should occur can also be beneficial for both the person responsible for data capture and people responsible for subsequent phases such as data cleaning.

Here is an example of data capture steps you might add to a student survey data collection SOP.

```
[Project Coordinator Name]

1. Download a CSV file with the following options selected
     - Use numeric values
     - Recode seen but unanswered questions as -99
     - Split multi-response fields into columns
2. Save the file in wave# -> student -> survey -> raw
     - Name the file "pn_w#_stu_svy_raw_YYYY-MM-DD.csv"
3. Add a README to the folder as needed to describe any
known issues with the data
     - Add a changelog as needed when new datasets are
downloaded
4. Open and review the file
     - Compare the variables in the file to the data
dictionary to ensure the file looks as expected
     - Compare the number of rows in the file to the
number of surveys tracked as complete in the participant
tracking database
          - If the numbers do not align, correct errors (or
note them in a README) and redownload the survey as
needed
5. Notify the data manager that the data is ready for
cleaning
```

12.2 Paper Data Capture

The most common method for capturing paper forms is manual entry. While capturing electronic data is fairly quick and straightforward, planning for and implementing paper data entry is much more involved. Similar to electronic data collection, you will want to start planning data entry long before your data is collected, and you will need to build your data entry tool before the data capture phase (e.g., when you are creating your data collection tools).

As you can imagine, manually entering data comes with the potential for many data quality issues. In developing a data entry process, it is important to implement quality assurance and control practices similar to those we discussed in Chapter 11.

1. Choose a quality data entry tool
2. Build your data entry form with the end in mind
3. Develop a data entry procedure

12.2.1 Choose a Quality Data Entry Tool

Unless you are entering data into a proprietary scoring tool associated with your instrument, you will need to choose a data entry tool. When choosing this tool, if you are already using a relational database for your participant tracking, it may make the most sense to use this same database for data entry so that data can be stored in one location and tables can be linked (e.g., REDCap, FileMaker, Microsoft Access). However, if you need to choose a new tool for data entry, the criteria for choosing one will be similar to those reviewed in Section 11.2.3. Considerations for project needs, security, costs, and data quality should all still be reviewed.

In addition to reviewing those criteria, it can also be very beneficial to use a tool that allows you to create entry forms (also called data entry screens), similar to the form we saw in Figure 10.9, rather than entering directly into a spreadsheet. Building a data entry screen that is laid out similar to the paper form can help reduce errors in data entry. Data that is entered into the form is then fed into a table that can be exported.

If however, you choose to use a spreadsheet program such as SPSS or Microsoft Excel for data entry, it is important to be aware of some of the limitations and possible issues with these tools including:

- Possible formatting issues
  - For example, Microsoft Excel formatting may cause errors in your data (e.g., dates get formatted as numeric, strings get formatted as dates, leading zeros get dropped from values).
- Potential to skip around
  - In a spreadsheet, the ability to click anywhere makes it very easy to enter data into the wrong cell or to skip cells completely (Eaker 2016). You may even delete or write over existing data on accident. It's also possible to incorrectly sort data resulting in errors (Reynolds, Schatschneider, and Logan 2022).

12.2.2 Build with the End in Mind

When you export or save a dataset from your data entry tool, it should meet all of our data organization rules (see Chapter 3), and all of the variables should be formatted as we have described in our data dictionary, including correct name, variable type, and allowable values. In order to accomplish that goal, you need to build your data entry screens, whether in a spreadsheet or form layout, following rules similar to those discussed in Section 11.2.4.

1. Include every item on your form.
   - For assessments consider entering both item-level data and summary variables derived in the field. Both types of variables can be beneficial to have in your final project datasets. Make sure that all of these variables are also accounted for in your data dictionary.
2. Make sure that your items are laid out in the same order that they appear on the paper form so that people entering the data can easily follow the flow (Reynolds, Schatschneider, and Logan 2022).
3. Using the annotated instrument we discussed in Section 11.2.4.2, name all of the items on your data entry screen to match the final item names (e.g., instead of `Q2` use the final name `tch_years`).
4. For quicker data entry, with less errors, allow people to enter the numeric values associated with response options on the annotated instrument rather than the text values (e.g., enter 1 rather than "strongly disagree"). Or if you prefer to use text values, build those as drop-down values, removing variation in entry.
5. Design the data entry tool to include both content and response validation.
   - Restrict data type, format, ranges, and values.
   - Do not allow people to skip over items.

Before using your data entry tool, you will want to pilot it for issues, just like we did for electronic data collection tools (see Section 11.2.4.1). Collect sample responses from team members and collect feedback on what did or did not work well for them while entering data. Then download, using your chosen download format, or simply review the data if it is already in its final format (e.g., Microsoft Excel). Check that the data looks as you expect it to and make edits to the entry tool as needed.

12.2.3 Develop a Data Entry Procedure

While building a reliable data entry tool is absolutely important in ensuring data quality, developing a clear and standard data entry process is even more important. Figure 12.4 shows a series of decisions to make regarding the data entry process.

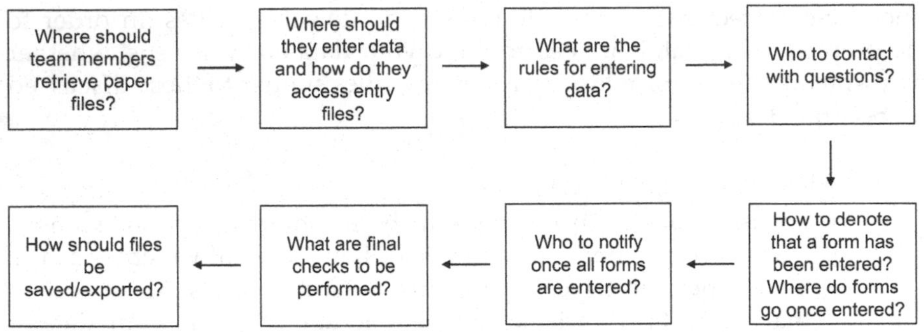

FIGURE 12.4
The flow of the decisions to make regarding the data entry process.

Let's walk through some of the decisions to be made.

1. Who will create data entry tools, who will enter data, and who will oversee this process?
2. Where will paper forms be stored and how should they be pulled?
3. Where will entry databases or spreadsheets be stored and how can they be accessed?
4. What are the data entry rules?
    - What values are entered for categorical variables (numeric values or text values)?
    - How should free-text be entered to prevent inconsistencies?
        - While adding data validation (e.g., only allowing numeric values) will help remove some inconsistencies, further rules may be needed depending on the items. Examples of rules include:
            - Enter decimals with a leading zero (e.g., *0.4*, not *.4*)
            - Enter "yes" values as "Y" (e.g., change any values of "y" or "yes" -> "Y")
            - Only enter numeric values for measurements (e.g., *5* not "5cm")
    - How should missing data be handled (e.g., fill with *-99*)?
    - What should be done if a team member comes across common data errors?
        - Someone who has circled more than one response to an item?
        - Someone who has written responses in the margin?

    - Someone who has written a value out of range or an unallowable response?
        - In general, I recommend to enter the exact values a participant has provided and rectify errors during the cleaning process. However, there may be reasons your team has to deal with certain errors during the entry process (e.g., your entry form is set up to only allow values *1–3* and a participant has provided a value of *4*).

5. How should team members denote that a form has been entered?
    - For example, staff can write their initials on a form after entry and then files can be stored in a new drawer or location to denote first or second entry.
6. What steps should be performed before moving on to the next phase of data cleaning?
    - Similar to the process in Section 12.1, it is imperative that whoever is overseeing the data entry process do a check of the data before handing it off for the next step of data cleaning.
    - Most importantly, check to see that the correct number of participants exists in the file compared to the number of participants with data marked as collected in your tracking database (i.e., no duplicate entries, no missing entries).
        - If data entry or data tracking errors exist, fix mistakes as needed before finalizing the entry process.
7. Last, what file type should be exported or saved (e.g., CSV, XLSX, SPSS), where should the files be stored, and how should they be named?
    - Store according to your storage rules (see Chapter 13) and name files according to your style guide (see Section 9.3).

NOTE

In general the data capture phase is a time to do only that, capture the data that is already collected. This is not a time to score, calculate, or add additional fields. This is a time to enter the items that are found on the form. Creating additional variables or performing further data quality checks will occur during the data cleaning phase.

An exception to this rule is if you collected an assessment that requires entry into a proprietary scoring program. Once data is entered, these tools often export a file that includes derived scores for the assessment, and these are still considered raw captured data sources.

12.2.3.1 Double Entry

Last, data quality can be further improved by double entering data. In their studies, Schmitt and Burchinal (2011) have found an error-rate between 5% and 10% when data are entered only once and that having a second person double-check data entry improves data quality. While there are several ways of double checking data including visual checking and read aloud methods, the double entry method has been shown to be the most reliable error-reducing technique (Barchard et al. 2020), ensuring that what is displayed on the paper form is what is entered into the database. A typical double entry process looks something like this.

1. A designated team member creates two identical entry forms. One person enters forms in the first entry screen; a different person enters forms in the second entry screen. Depending on your tool this might be two separate files, two separate tabs in a spreadsheet, or two separate tables or forms in a database.
    - It is important here that the second entry is completed by a different person so that systematic errors that are created by one person's interpretation of information are not repeated across files.
2. When both entries are complete, a system is used to check for inconsistencies across datasets.
    - This system varies across tools. Some tools have built in systems for checking for errors across entry screens. Other tools may require you to build your own system (e.g., write formulas to compare cells or draft syntax to compare spreadsheets). Ultimately, once those comparisons are done, you should have a report that tells you where errors exist across the two forms.
3. Using the information gained from comparing entry screens, a designated team member makes corrections (Yenni et al. 2019). This involves pulling out the original paper forms and seeing what the correct value is for each error.
    - There are varying ways you can make corrections at this point. You can make corrections just to one form, you can make corrections to both forms, or you can make corrections in a third, new form that contains all of the correct data. Different tools will handle this in different ways.
    - However, if you are creating your own system, consider making corrections in both forms. In this way, you make a correction to whichever entry file has the error. Once all corrections are made, you can run your comparison system again, which will let you

know if all errors have been corrected. Once all errors are fixed, you can choose either file to be your "master" raw data file.

The following is one example of what this process could look like.

1. Data is entered into two spreadsheets following quality assurance and control procedures. Then, both files are imported into R.
2. A function from the `diffdf` package[1] (Gower-Page and Martin 2020) is run to check for errors and a report is returned (see Figure 12.5).
   - You can see that it identifies an error in our `t_mast1` variable. For `tch_id` = 236, entry file 1 (*BASE*) has a different value than entry file 2 (*COMPARE*).
3. Paper forms are checked to see what the true reported answer is and then the value is corrected in the corresponding entry file. If the value is incorrect in both entry files, both files are corrected.
4. Updated files are imported back into R and the comparison system is run again to ensure no more errors exist.

Depending on the amount of data that is collected this can be a time-consuming process. Double data entry is a matter of weighing costs and benefits. While double entering all of your data is the best way to reduce data errors, the cost

```
Differences found between the objects!

A summary is given below.

Not all Values Compared Equal
All rows are shown in table below

  =============================
   Variable  No of Differences
  -----------------------------
   t_mast1           1
  -----------------------------


All rows are shown in table below

  =================================
   VARIABLE  tch_id  BASE  COMPARE
  ---------------------------------
   t_mast1    236     2      4
  ---------------------------------
```

FIGURE 12.5
A report displaying differences between two entry files.

of double entering all of your data might be too high, and you may decide to only double enter a portion of your data and gain a smaller benefit.

12.2.4 Documenting Paper Data Capture

Whatever your decisions are throughout this process, document them in an SOP and assign team members for each step. This includes assigning someone to create entry files, oversee data entry, create a double entry comparison system, conduct the comparison, make corrections, and do the final checks before handing the data off. Make sure to train your team on this system so that it is implemented consistently. For any types of technical processes, in addition to in-person training sessions, it can also be helpful to record and share videos covering procedures that team members can review as often as necessary.

12.2.5 Scanning Forms

It is possible that you may collect paper data using forms which can be scanned and converted automatically into a machine-readable dataset. Depending on whether your team is personally doing the scanning or whether an external company captures the data, this has the potential to save you time and energy compared to a manual data entry process. These may also have the potential to be less error-prone than manual entry, yet this process is still not error-free and caution should be taken when capturing this data (Jørgensen and Karlsmose 1998). It is still important to do data checks to ensure that the correct values were recorded in the electronic file.

12.3 Extant Data

It is common in education research to also capture external supplemental data sources to either link to your original data sources or to describe information about your sample. The process for capturing this externally collected data will vary widely depending on the source. Furthermore, the quality and usability of the data can also vary greatly. In this section we are going to review some practices that will help you acquire better, more interpretable data. We will divide this discussion between two types of data sources—non-public and public.

12.3.1 Non-Public Data Sources

Non-public, or restricted-use, data sources are files that cannot be directly accessed from a public website (e.g., student education records, statewide longitudinal data systems). These data are typically individual-level and

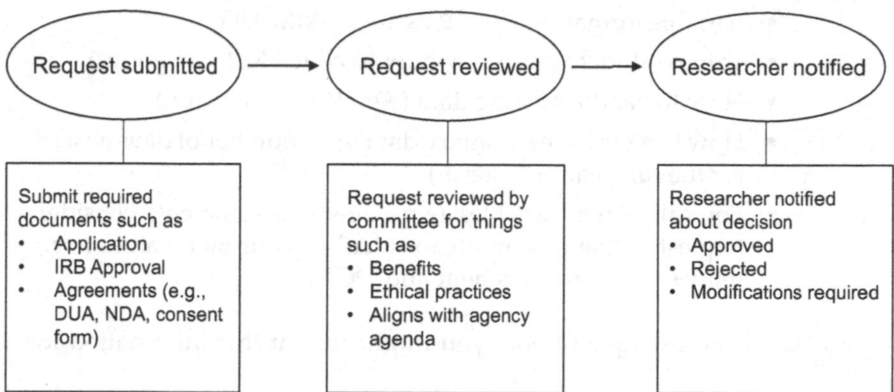

FIGURE 12.6

Example non-public confidential data request process.

may contain sensitive, usually identifiable, information or a combination of variables that could enable identification. Acquiring these sources usually involves a data request process. Every data request system will be different, but Figure 12.6 provides an example of what this process might look like. If conducting research in schools, requesting student education records may already be a part of your research request process (see Section 4.3.3). It is important to begin looking into this during your planning phase to understand when the request should be initiated.

If not already included in the provider's data request process, it is important to add quality assurance to the process by sharing the following information with a provider.

1. The periods you are requesting data for (e.g., 2023–2024 school year)
2. The variables you are requesting
    - If you plan to link data, make sure your list includes a unique identifier that allows you to link the external data to your existing original data.
3. Variable details
    - Giving the provider details regarding your preferred variable format can help standardize inputs across sources (e.g., multiple school districts) and reduce confusion (Feeney et al. 2021). Some external data providers will be able to accommodate your specific requests, others will not. If the provider is able to accommodate your specific needs, the following types of information can be helpful to provide:
        - Variable type (e.g., numeric, text)

- Variable formats (e.g., DOB as YYYY-MM-DD)
- Value coding (e.g., specify how to code FRPL categories)
- How to handle missing data (e.g., leave cell blank)
- How to aggregate summary data (e.g., number of days absent for the full year **or** by term)
- For calculated variables (e.g., age at assessment) consider requesting the raw inputs to calculate your own values (e.g., request date of assessment and DOB)

Figure 12.7 is an example of how you might format this information for a provider.

You will also want to make sure you acquire the following information:

1. How and when data will be shared
   - How many data files will be provided and what will each file will contain (e.g., enrollment file, assessment file, attendance file)?
   - What file formats will be provided (e.g., CSV files)?
   - When will data be shared (i.e., timeline)?
   - How will data will be shared (e.g., email, drop in a secure folder)?
     - If the data contain identifiable information, make sure to use a secure file transfer method (see Chapter 13).
2. Who are the points of contact
   - Not only do you need contact information for acquiring the data, you also need to know who to contact for any questions or concerns that come up after the data is received.
3. Any documentation to accompany your file
   - Receiving data dictionaries or codebooks along with your data will be vital in allowing you to correctly interpret variables. This is especially important when observing variations in how variables are measured across sites or even within sites across time (e.g., a test score is measured differently in a subsequent year).
   - If documentation does not exist, provide the data provider with a form to complete that allows them to enter relevant, variable information (see Figure 12.8 for an example).
     - What each variable represents
     - Allowable variable ranges
     - Categorical code definitions
     - How each variable is captured or calculated (e.g., hand-entered)
     - The universe for each variable (e.g., grades 3–5)

Variable Requested	Variable Type	Description	Categorical Codes
State unique identifier	Numeric	Student state unique identifier	NA
Free or reduced-price lunch	Numeric	Free or reduced price lunch status for 23-23 school year	0 = none | 1 = reduced | 2 = free
Grade level	Numeric	Grade level for 23-24 school year	0 = kindergarten | 1 = first | 2 = second | 3 = third | 4 = fourth | 5 = fifth
Date of birth	Date (YYYY-MM-DD)	Student date of birth	NA
State testing date - math	Date (YYYY-MM-DD)	Math MAP testing date for 23-24 school year	NA
State testing score - math	Numeric	Math MAP score for 23-24 school year	NA
Number of in-school suspensions	Numeric	Total number of in-school suspensions for 23-24 school year	NA

FIGURE 12.7
Example variable request for an external data provider.

Variable Requested	Variable Name	Description	Categorical Codes	Data Quality Concerns
State unique identifier				
Free or reduced-price lunch				
Grade level				
Date of birth				
State testing date - math				
State testing score - math				
Number of in-school suspensions				

FIGURE 12.8
Sample documentation form for an external data provider to complete.

  - Any data quality concerns about any of the variables
- If you receive new exports each year, make sure to request documentation each year. It is possible that the way variables are collected or recorded change over time.

Once you receive external files, make sure to perform quality control checks before saving them and preparing for further data cleaning.

1. Review files to ensure all participants are accounted for and all requested variables are included.
   - Track the information received in your participant tracking database (i.e., track school record received for `stu_id` = 1234). This will help you determine if you have duplicates or missing cases.
   - Compare the variables in the data to the variables from your request. Is everything accounted for?
   - Reach back out to your contact for help as needed.
2. Make sure you have all necessary information to correctly interpret the data (i.e., any required documentation).
3. Last, store files in a manner that adheres to applicable agreements and name the files according to your style guide rules.

> **NOTE**
>
> When working with external datasets, it is possible to encounter inconsistencies across data sources (e.g., a student is shown in a different school across two files), as well as duplicate records within a data source (e.g., a student has two state reading assessment scores) (Levesque, Fitzgerald, and Pfeiffer 2015). These anomalies can happen due to human error or due to circumstances such as student mobility. While you may be able to work with your data provider to solve some data issues, for others it may be important for you to develop and document your own data management rules that you consistently apply to your external data sources during the data cleaning phase (e.g., if duplicate assessment records exist, the earliest assessment date is used).

12.3.2 Public Data Sources

Publicly available data sources are typically aggregated (i.e., state-, district-, or school-level) or de-identified individual-level datasets that are available through various agencies such as state departments of education or federal agencies. These datasets are often extracted by downloading a file, although some organizations may have more sophisticated API capabilities. The quality of these datasets may vary. A few tips for working with publicly available datasets are:

1. Extract the data early on in your project.
   - Even if it is not the most up to date data that you need, it's important to get a sense early on for what the data looks like (e.g., what variables are included, what file types data are stored in, how the files are structured). This helps you prepare for future data wrangling needs.
2. Find the associated documentation and read it thoroughly. Types of documentation to look for are:
   - Data dictionaries or codebooks
     - These documents will help you interpret and use variables correctly.
   - Changelogs
     - Public data sources are constantly updating (e.g., new data is acquired, errors are found). It's important to understand what version of the data you are working with.

- Data quality documentation
  - This documentation helps make you aware of any known issues in the data.

3. Do not hesitate to reach out for help.
   - Typically, the site will include contact information for questions. Never hesitate to reach out to that contact if there is something you do not understand in the data.
4. If extracting data across states (e.g., Missouri Department of Elementary and Secondary Education and Oklahoma State Department of Education), be aware that the information may not be easily comparable. While you may find that some states use similar standards, it is common for states to collect and store data in different ways (e.g., different state assessments, different ways of reporting enrollment). Depending on your data needs, it may be better to use a data source that aggregates information across states. Examples of such data sources include the National Center for Education Statistics' Common Core of Data (https://nces.ed.gov/ccd/) or EDFacts (https://www2.ed.gov/about/inits/ed/edfacts/index.html). If you are needing to use multiple data sources, other tools, such as the Urban Institute's Education Data Explorer (https://educationdata.urban.org/data-explorer), have even harmonized variables and documentation across several federal government datasets, allowing researchers to access multiple data sources in a single site.
5. Once files are extracted, store them according to your data storage plans (see Chapter 13) and name them according to your style guide rules.

12.3.3 Documenting External Data Capture

Make sure every step of your external data capture process is documented in the appropriate locations (e.g., SOP, research protocol). Responsibilities will need to be assigned throughout this process, from investigating data request procedures, to communicating with providers, to acquiring data files. As mentioned in Section 8.4.1.2, you will also want to make sure these data sources are documented in a data dictionary. It is possible your data source will come with its own data dictionary, but for continuity, you may need to reformat that document to match how your other project data dictionaries are formatted. If your data source does not come with a data dictionary, you will want to create one for this source. This will be useful not only for data cleaning and validation purposes, but also for data sharing purposes later on.

Note

1 https://osf.io/saut6

13

Data Storage and Security

As you begin to capture data, it is important to have a well-planned structure for securely storing and working with that data during an active study (see Figure 13.1). Not only do you need a plan for storing data files, but you also need a plan for storing other project files (e.g., meeting notes, documentation, participant tracking databases). Your team should implement this structure early on so that files are stored consistently and securely for the entire project, not just once the data collection life cycle begins. There are several goals to keep in mind when setting up your file storage and security system for an active project.

1. File security: Ensuring that your files are not lost, corrupted, or edited unexpectedly.
2. Protecting confidentiality: Making sure that sensitive information is not seen or accessed by unauthorized individuals.
3. Accessibility and usability of files: Making sure that your team can easily find files and that they are able to understand what the files contain.

13.1 Planning Short-Term Data Storage

When planning a storage and security process, for data files in particular, it is important to gather all relevant information before making a plan. A typical process for developing a plan may begin like this:

1. Review what data needs to be stored and how often.
    - Use documents such as your data sources catalog (see Section 5.3) and your data collection timeline (see Section 8.2.6) to better understand your data storage needs.

DOI: 10.1201/9781032622835-13

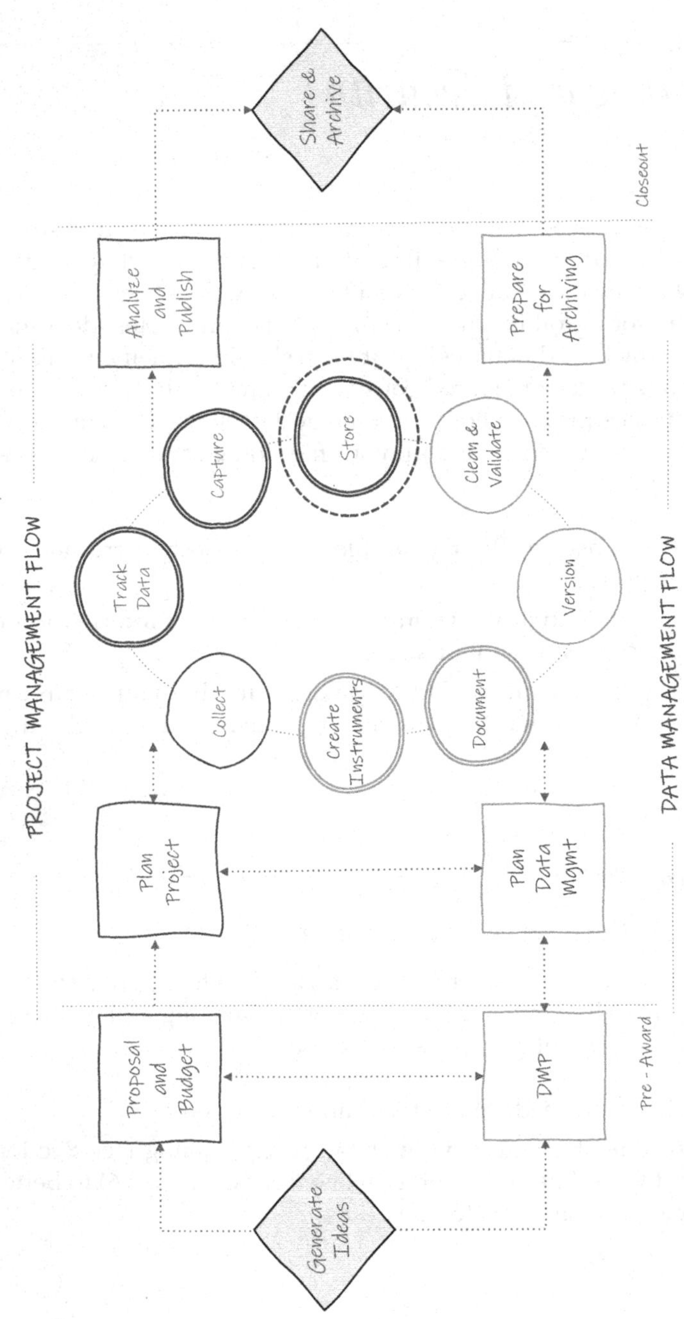

FIGURE 13.1
Data storage in the research project life cycle.

2. Take an inventory of what data storage solutions are available to you.
   - In terms of electronic data, institutions have different licenses or partnerships with varying software companies and they may approve and not approve different tools (e.g., Dropbox, SharePoint, Box, Google Drive).
3. Consider compliance.
   - Make note of any data storage laws, policies, or agreements that your data is subject to (e.g., IRB policies, data sharing agreements, funder policies).
4. Review classification levels.
   - Review each data source's classification level (see Section 4.2) to ensure that you are choosing storage solutions that are appropriate for the sensitivity level your data.

This process should help narrow down your data storage solutions for each data source. However, from there a series of decisions need to be made depending on the type of data you are working with, paper (e.g., a paper consent form) or electronic (e.g., a CSV file of questionnaire data, a Microsoft Access participant tracking database). The remainder of this section will review a series of decisions to make for each type of data, as well as provide some best practices along the way.

13.1.1 Electronic Data

Once you have reviewed all relevant information from Section 13.1, several more decisions will need to be made when choosing and setting up your structures for storing and securely working with electronic data.

1. Review additional criteria.
   - After narrowing down storage solutions based on available tools (e.g., cloud storage, institution network drive, personal device) that meet your compliance needs, electronic data storage locations can be further narrowed based on other criteria.
     - Versioning availability: While manual version control is beneficial for major changes, it is very helpful to store your files in a location that has automated file versioning as a fail-safe in case of accidents such as unintended overwriting of files.
     - Size of the storage space: You will need to make sure your storage contains enough space for your files.

    - Consider how many files you will be storing, as well as the amount of information stored in each file (e.g., number of rows and columns in each dataset). The file types you use (e.g., CSV, XLSX, SPSS) also impact the size of a file.
  - Comfort level of your team: It is helpful to choose a storage space that your team is comfortable working in or that you have the ability to train them in how to use it.
  - Accessibility: Consider the accessibility of your storage location for users (e.g., how staff access the location off-site), as well as the compatibility with different operating system.
  - Collaboration: Consider how the storage method handles multi-user editing of files.
  - File sharing: It can be very beneficial to use a storage platform that allows file sharing through links, rather than sharing the actual file. Not only does this better control file access, it also reduces burden and confusion. As an example, if small updates are made to a file after sharing, rather than having to send an updated version, those changes are reflected in the linked file.
  - Costs: Consider if there are any costs associated with your potential storage solutions.

2. Choose a final storage location.
   - While you may be allowed to store files in different locations depending on their sensitivity level, a more effective solution is to create a collaborative research environment (UK Data Service 2023). To do this, designate the highest level of security needed (e.g., an institution network drive), and keep all, or as many as possible, project-related files stored in that same location, assigning access to files and folders as needed. Keeping all files located in a central, consistent location often provides the benefit of data security (e.g., user access controls, not having different versions of documents on different computers) as well as accessibility (e.g., team members can find documents).
3. Design your folder structure according to your style guide.
   - Following your style guide (see Section 9.2), create a folder structure **before** team members begin storing files so that they are stored consistently.
   - Importantly, make sure to set up your structure in a way that stores identifiable information (e.g., participant tracking database, consent forms) separate from your research study data.

4. Set up additional security systems.
   - Data backups.
     - It is important to regularly backup up your data. Consider using something similar to the 3-2-1 rule, keeping three copies of your data, on two different types of storage media, in more than one location to guard against data loss (Briney 2015; UK Data Service 2023). Talk with your institution IT department for help with setting up this system.
   - User access.
     - Assign user access to folders and files based on sensitivity levels, quality control needs, and applicable policies, agreements, or plans.
5. Designate rules for working securely with data.
   - Complete required trainings (e.g., CITI trainings, IT training, internal training).
   - Consistently name folders and files according to the style guide (see Sections 9.2 and 9.3).
   - Do not keep copies of files.
     - Outside of making data backups, do not keep copies of files in different folders. This opens the door for edits being made to one copy and not the other. If this happens, team members may be working with different versions of files. If you want to have a copy of a file in more than one location (e.g., an SOP in the "documentation" folder and the "project_coordination" folder), some storage systems allow you to link to documents from other location (i.e., the "project_coordination" folder contains a link to an SOP in the "documentation" folder).
   - Secure your devices (O'Toole et al. 2018; Princeton University 2023a).
     - Choose safe passwords to protect devices.
     - Do not leave devices open and unattended when working in the field.
     - Have protection on your devices (e.g., up-to-date antivirus software, firewall, encryption).
     - When working remotely, use password-protected Wi-Fi and use secure connections (e.g., VPN, 2FA) when working with data files.
     - Any files stored on detachable media (e.g., external hard drives, CDs, flash drives) should typically be stored behind

    two locks when not in use (e.g., a locked file cabinet in a locked storage room).
- Securely transmit data files.
    - When transmitting data, either internally or externally, it is important that you use secure methods, especially when data contain PII. As a general rule, no moderate or highly sensitive data should be transmitted via email. Use a secure, institution-approved, file transfer method that includes encryption.

13.1.2 Paper Data

Working with paper data involves reviewing another set of decisions when planning for data storage and security.

1. Choose a final storage location.
    - After reviewing available locations as well as applicable laws, policies, and agreements, you will want to consider additional criteria such as accessibility of the storage site, your physical storage size needs, storage costs, and the security of the location. Most commonly required for any files containing PII is to store them behind two locks.
2. Consistently structure your file cabinets and folders.
    - While you may not have a style guide created for organizing physical folders and files, it is still important to consistently structure and name them for clarity. As an example, organize drawers by time period (e.g., wave 1), and further organize folders by data source (e.g., student survey).
3. Securely work with files.
    - As discussed in Sections 11.3.1 and 12.2.3, as team members work with files, it is important that staff understand the rules and process for returning files back to the designated storage location when not in use (i.e., no files left on desks).

13.1.3 Oversight

Whether working with electronic or paper data, make sure to assign responsibilities to team members for tasks such as creating electronic directory structures or physical folder structures, adding and removing storage access, overseeing data backups, and monitoring training compliance. Without oversight of these processes, it is easy for errors to occur.

13.2 Documentation and Dissemination

After you make a plan for short-term data storage, that plan should be added to all necessary documentation (e.g., DMP, research protocol, informed consent forms, SOPs). Once your plan has been approved, it is important to not deviate from that plan unless your revisions have also been approved. This is especially important in the case of agreements (e.g., informed consent, data sharing agreements), where it is important to honor any terms laid out in those agreements.

Last, all information needs to be disseminated to team members to ensure fidelity to your data storage and security plan. Add pertinent information to documents that staff are required to review (i.e., team data security policy or onboarding/offboarding checklists) to ensure the information is reviewed. In addition, make sure to embed this information into any team- or project-related staff training.

14

Data Cleaning

Even with the most well-designed data collection and capture efforts, data still require at least some additional processing before it is in a format that you will feel confident in sharing. What is done in the data processing, or data cleaning phase, will largely depend on the planned transformations for your data, as well as the level of quality assurance and control processes implemented during collection and capture. In this chapter we will review some standard data cleaning steps that should be considered for every education research project. What is important to emphasize here, is that if you are collecting longitudinal data, data cleaning needs to happen every wave of data collection (see Figure 14.1). Once a wave of data has been collected and captured and the raw data has been stored, your data cleaning process should begin. In a best-case scenario, the data cleaning is wrapped up before your next wave of data collection. Cleaning data each wave, as opposed to waiting until the end of your project, has two large benefits.

1. Allows you to catch errors early on and fix them.
   - While cleaning your data you may find that all data is missing unexpectedly for one of your variables, or that values are incorrectly coded, or that you forgot to restrict the input type. If you are cleaning data each wave, you are able to then correct any errors in your instrument in order to collect better data next round.
2. Data is ready when you need it.
   - Proposal, report, and publication deadlines come up fast. As various needs arise, rather than having to first take time to clean your data, or waiting for someone on your team to clean it, data will always be available for use because it is cleaned on a regularly occurring schedule.

14.1 Data Cleaning for Data Sharing

Data cleaning is the process of organizing and transforming raw data into a dataset that can be easily accessed and analyzed. Data cleaning can essentially

DOI: 10.1201/9781032622835-14

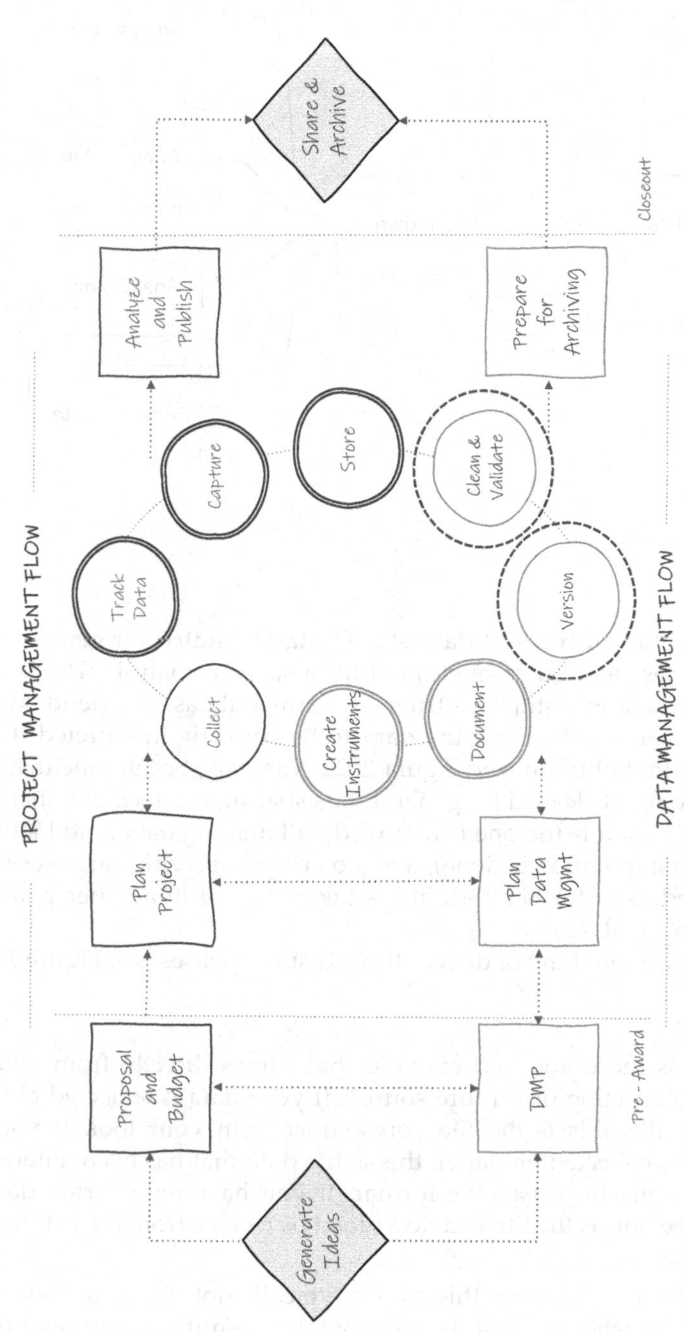

FIGURE 14.1
Data cleaning in the research project life cycle.

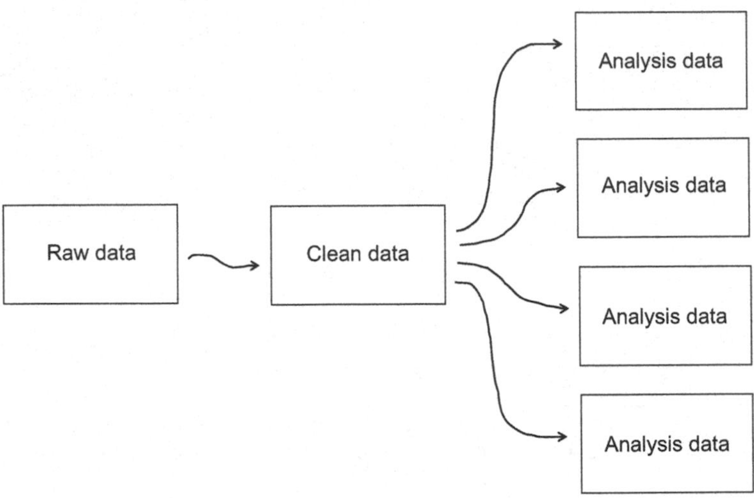

FIGURE 14.2
The three phases of data.

result in two different types of datasets; a dataset curated for general data sharing purposes, and a dataset cleaned for a specific analysis. The former means that the dataset is still in its true, raw form, but has been de-identified and minimally altered to allow the data to be correctly interpreted (Cook et al. 2021; Neild, Robinson, and Agufa 2022; Van Dijk, Schatschneider, and Hart 2021). A dataset cleaned for general data sharing means that it includes the entire study sample (no one is removed), all missing data is still labeled as missing (no imputation is done), and no analysis-specific variables have been calculated. Any further cleaning is taken care of in another phase of cleaning during analyses.

Ultimately, you can think of data in three distinct phases (see Figure 14.2).

1. Raw data
    - This is the untouched raw file that comes directly from your data collection or capture source. If your data is collected electronically, this is the file you extract from your tool. If your data is collected on paper, this is the data that has been entered into a machine-readable format. If you have an external data source, this is the file you download or receive from the external provider.
    - In education research this data is typically not shared outside of the research team as it usually contains identifiable information and often needs further wrangling to be decipherable by an end user.

2. The general clean study data
    - This is the dataset that you will publicly share and is the one we will be discussing in this chapter.
3. Your analytic data
    - This dataset is created from the general clean dataset (either by your team or by other researchers), but is further altered for a specific analysis (Reynolds, Schatschneider, and Logan 2022). This dataset will typically also be publicly shared in a repository at the time of publication to allow for replication of the associated analysis. Since this dataset is analysis specific, we will not discuss this type of data cleaning in this book.

14.2 Data Quality Criteria

Before cleaning our data, we need to have a shared understanding for what we expect our data to look like once it is cleaned. Adhering to common standards for data quality allows our data to be consistently cleaned and organized within and across projects. There are several data quality criteria that are commonly agreed upon (DeCoster 2023; Elgabry 2019; Schmidt et al. 2021; Towse, Ellis, and Towse 2021; Van Bochove, Alper, and Gu 2023). Upon cleaning your data for general data sharing, your data should meet the following criteria.

1. Complete
    - The number of rows in your dataset should match the number of completed forms tracked in your participant tracking database. This means that all forms that you collected have been captured (either entered or retrieved). It also means that you have removed all extraneous data that doesn't belong (e.g., duplicates, participants who aren't in the final sample).
    - The number of columns in your data match the number of variables you have in your data dictionary (i.e., no variables were accidentally dropped). Similarly, there should be no unexpected missing data for variables (i.e., if the data was collected, it should exist in your dataset).
2. Valid
    - Variables conform to the constraints that you have laid out in your data dictionary (e.g., variable types, allowable variable values and ranges, item-level missingness aligns with variable universe rules and defined skip patterns)

3. Accurate
    - Oftentimes there is no way to know whether a value is true or not.
        - However, it is possible to use your implicit knowledge of a participant or a data source (i.e., ghost knowledge) (Boykis 2021) to determine if values are inaccurate (e.g., a value exists for a school where you know data was not collected that wave).
        - It is also possible to check for alignment of variable values within and across sources to determine accuracy.
            - For example, if a student is in 2nd grade, their teacher ID should be associated with a 2nd-grade teacher. Or, a date of birth collected from a student survey should match date of birth collected from a school district.
4. Consistent
    - Variable values are consistently measured, formatted, or categorized within a column (e.g., all values of survey date are formatted as YYYY-MM-DD).
    - Across repeated collections of the same form, all variables are consistently named, measured, formatted, or coded as well (e.g., free/reduced priced lunch is consistently named and coded using the same code/label pair across all cohorts).
5. De-identified
    - If confidentiality is promised to participants, data needs to be de-identified. At the early phases of data cleaning, this simply means that all direct identifiers (see Chapter 4) are removed from the data and replaced with study codes (i.e., participant unique identifier). Before publicly sharing data, additional work may be required to remove indirect identifiers as well and we will discuss this more in Chapter 16.
6. Interpretable
    - Variables are named to match your data dictionary and those names are both human- and machine-readable (see Section 9.4). As needed, variable and value labels are added as embedded metadata to aid in interpretation.
7. Analyzable
    - The dataset is in a rectangular (rows and columns), machine-readable format and adheres to basic data organization rules (see Section 3.2).

14.3 Data Cleaning Checklist

Recall from Section 8.3.3, that it is helpful to write a data cleaning plan, for each dataset in your data sources catalog (see Section 8.2.2), before you begin cleaning your raw data. Writing this plan early on allows you to get feedback on your planned alterations, and also provides structure to your cleaning process, preventing you from meandering and potentially forgetting important steps. This plan does not need to be overly detailed, but it should include actionable steps to walk through when cleaning your data.

In many ways, writing this data cleaning plan will be a very personalized process. The steps needed to wrangle your raw data into a quality dataset will vary depending on what is happening in your specific raw data file. However, to produce datasets that consistently meet the data quality standards discussed in Section 14.2, it can be helpful to follow a standardized checklist of data cleaning steps (see Figure 14.3). These steps, although very general here, once elaborated on in your data cleaning plan, for your specific data source, can help you produce a dataset that meets our data quality standards. Following this checklist helps to ensure that data is cleaned in a consistent and standardized manner within and across projects.

As you write your data cleaning plan, you can add the checklist steps that are relevant to your data and remove the steps that are not relevant. The order of the steps is fluid and can be moved around as needed. There are two exceptions to this. First, accessing your raw data will always be number one of course, and the most important rule here is to never work directly in the raw data file (Borer et al. 2009; Broman and Woo 2018). Either make a copy of the file or connect to your raw file in other ways where you are not directly editing the file. Your raw data file is your single source of truth for that data source. If you make errors in your data cleaning process, you should always be able to go back to your raw data to start over again if you need to. Second, reviewing your data should always be step number two. Waiting to review your data

- ☐ Access raw data
- ☐ Review raw data
- ☐ Find missing data
- ☐ Adjust the sample
- ☐ De-identify data
- ☐ Drop irrelevant columns
- ☐ Split columns
- ☐ Rename variables
- ☐ Normalize variables
- ☐ Standardize variables
- ☐ Update variable types
- ☐ Recode variables
- ☐ Construct new variables
- ☐ Add missing values
- ☐ Add metadata
- ☐ Validate data
- ☐ Join data
- ☐ Reshape data
- ☐ Save clean data

FIGURE 14.3
Data cleaning checklist.

until after you've started cleaning means that you may waste hours of time cleaning data only to learn later that participants are missing, your data is not organized as expected, or even that you are working with the wrong file.

14.3.1 Checklist Steps

Let's review what each step specifically involves so that as you write your data cleaning plan, you are able to determine which steps are relevant to cleaning your specific data source.

1. Access your raw data.
    - Depending on how you plan to clean data (e.g., using code or manual cleaning), as well as the programs you plan to use for data cleaning (e.g., Excel, SPSS, R, Stata), this may look like reading a file into a statistical program, or it may look like making a copy of your raw data file and renaming it "clean". Either way, just ensure that you are not making edits to your raw data file.
2. Do a basic review of your raw data (see Figure 14.4).
    - Check the rows in your data
        - Does the number of cases in your data match the number of tracked forms in your participant tracking database? Are there any duplicates? Is anyone missing?
    - Check the columns in your data
        - Does the number of variables in your data dictionary match the number of variables in your dataset? Remember we are

Data dictionary

var_name	label	type	values
tch_id	Teacher study ID	numeric	200 – 300
t_sub_geo	Do you teach geometry?	numeric	1 = yes | 0 = no
t_sub_alg	Do you teach algebra?	numeric	1 = yes | 0 = no

Participant tracking database

tch_id	svy_complete
236	yes
237	no
238	yes

Raw data

tch_id	t_sub_geo	t_sub_alg
236	0	1
238	1	1

FIGURE 14.4
Reviewing rows and columns in a raw data file.

only looking for variables that are captured directly from our source (i.e., not variables that are derived during this cleaning process).

- Are the variable types and values as expected?

> **NOTE**
>
> Before starting the review process, you may need to do some pre-work to put data into an analyzable format (e.g., if your second row of data is variable labels, you will want to drop that second row so that you are only left with variable names in the first row and values associated with each variable in all remaining cells). Without doing this work first, it will be difficult to accurately review your data.

3. Find missing data.
    - Find missing cases.
        - If cases are marked as complete in your tracking database but their data is missing, investigate the error. Was a form incorrectly tracked in your tracking database? Was a form not entered during the data capture phase?
            - If there is an error in your tracking database, fix the error at this time.
            - Otherwise, search for missing forms, add them to your raw data, and start again at step number 1 of your data cleaning process.
    - Find missing variables.
        - If you are missing any variables, investigate the error. Was a variable incorrectly added to your data dictionary? Or, was a variable somehow dropped in the data capture process or in our data import or file copying process?
            - Fix the error in the appropriate location and then start again at step number 1.
4. Adjust the sample.
    - Remove duplicate cases.
        - First, make sure your duplicates are true duplicates (i.e., not incorrectly entered identifiers or data entry errors).
            - If the error comes from the data entry process (e.g., a team member entered an ID incorrectly, a team member

entered a form twice), it can be helpful to go back to your data entry files and fix those errors at the source. This way your raw data always reflects the true values reported by a participant, improving your ability to track data lineage. Then export new files and start the data cleaning process over again at step 1.

    - If the error comes directly from a data collection instrument (e.g., a participant entered an incorrect ID into the web-based form), those errors can be corrected now, in the data cleaning process.

  - If you have true duplicates (i.e., a participant completed a form more than once), duplicates will need to be removed. Follow the decisions written in your documentation (e.g., research protocol, SOP) to ensure you are removing duplicates consistently (e.g., always keep the first complete record of a form).

- Remove any participants who are not part of your final sample (i.e., did not meet inclusion criteria).

NOTE

In the special case where you purposefully collect duplicate observations on a participant (i.e., for reliability purposes), you will only want to keep one row per participant in your final study dataset. Again, a decision rule will need to be added to documentation so duplicates are dealt with consistently (e.g., always keep the primary observer's record).

5. De-identify data.
   - If confidentiality was promised to participants, you will need to de-identify your data. If your data does not already contain your assigned study IDs, replace all direct identifiers (e.g., names, emails) in your data with study IDs using a roster from your participant tracking database. This process creates coded data (see Chapter 4 for a refresher).
   - At this point we are focusing on removing direct identifiers only, but in Chapter 16, we will also discuss dealing with indirect identifiers before publicly sharing your data.

Raw data

first_name	last_name	t_stress1	t_stress2	t_stress3
Jana	Clark	1	3	4
Dennis	Nguyen	2	5	1
Kate	Malone	4	2	2

Roster file

first_name	last_name	tch_id
Jana	Clark	236
Dennis	Nguyen	237
Kate	Malone	238

Coded data

tch_id	t_stress1	t_stress2	t_stress3
236	1	3	4
237	2	5	1
238	4	2	2

FIGURE 14.5
Process of creating a de-identified dataset.

- Figure 14.5 shows what this data de-identification process might look like (O'Toole et al. 2018). In order to create a coded data file, we horizontally join our "raw data" file with our "roster file" using our unique identifiers (which in this case are `first_name` and `last_name`), and then drop our identifying variables.
  - I want to emphasize the importance of using a join in your program of choice, as opposed to replacing names with IDs by hand entering identifiers. If at all possible, you want to completely avoid hand entry of study IDs because it is error-prone and can lead to many mistakes.

NOTE

At this point in data cleaning, if your data contain open-text responses, you should be reviewing free text for identifiable information as well. Remove any instance of names and replace with a placeholder such as "<name>" to indicate the information was redacted.

6. Drop any irrelevant columns.
   - Here you can think of examples such as the metadata collected by a survey platform (e.g., survey duration). These columns may be irrelevant to your study and cause clutter in your final dataset.

7. Split columns as needed (see Figure 14.6).
    - As discussed in Section 3.2, a variable should only collect one piece of information. Here you will split one variable into multiple variables so that only one thing is measured per variable.
8. Rename variables.
    - Rename variables to correspond with the names provided in your data dictionary.
9. Normalize variables (see Figure 14.7).
    - Here, the term normalize is used to summarize the process of returning a variable to its normal, or expected, state.
    - Compare the variable types in your raw data to the variable types you expected in your data dictionary. Do they align? If no, why?
        - As an example, it may be that you need to remove unexpected characters such as $ or % that are preventing a variable from being a numeric type. Or it could be accidentally inserted white space or letters in your variable.

subject
algebra, geometry, calculus
geometry
algebra, geometry

→

sub_algebra	sub_geometry	sub_calculus
yes	yes	yes
no	yes	no
yes	yes	no

FIGURE 14.6
Splitting one column into multiple columns.

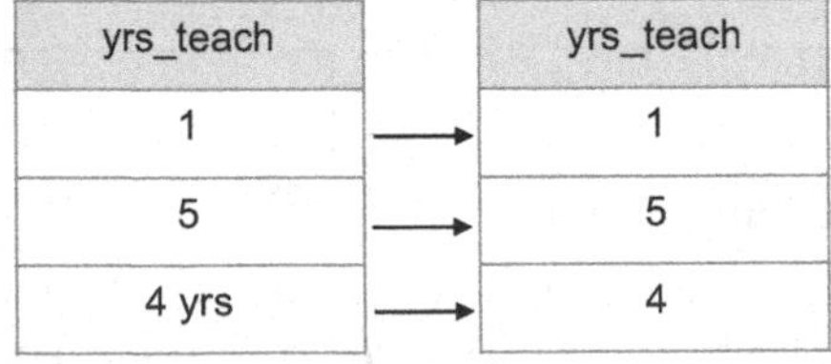

yrs_teach
1
5
4 yrs

→

yrs_teach
1
5
4

FIGURE 14.7
Normalizing a variable.

10. Standardize variables.
    - Here, I am using term standardize to convey the process of checking for consistency.
    - Are columns consistently measured, categorized, and formatted according to your data dictionary? If no, they need to be standardized.
        - This may involve rescaling variables (e.g., age measured in months in wave 1 and age measured in years in wave 2 would need to be rescaled).
        - This may mean updating a variable format (e.g., converting to a consistent date format).
        - Or it may mean collapsing categories of free text categorical variables (e.g., "m" | "M" | "male" = "male").

> **NOTE**
>
> In the case of Figure 14.8, this kind of standardization needs to happen before you can perform steps such as joining on names for de-identification purposes. Keys need to be standardized across files before linking can occur.

11. Update variable types.
    - After normalizing and standardizing variables, you can now convert any variable types that do not match the types you've listed in your data dictionary (e.g., convert a string to numeric)

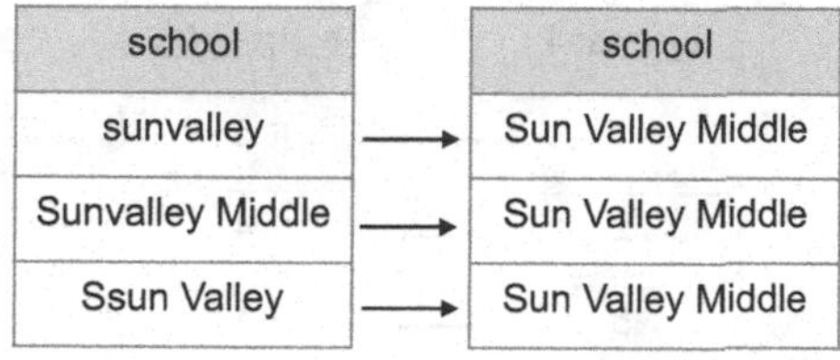

FIGURE 14.8
Standardizing a variable.

NOTE

It's important to normalize before updating your variable types. Updating your variable types before normalizing could result in lost data (e.g., converting a character column to numeric, when the column still contains cells with character values, will often result in those cells being recoded to missing).

12. Recode variables.
    - If your categorical value codes (see Section 9.5) do not match your data dictionary, now is the time to recode those (e.g., you expected "no" = 1, but the data exported as "no" = 14)
    - As discussed in Section 3.2, this also includes recoding implicit values, explicitly (e.g., if missing values are implied to be *0*, recode them to *0*)
    - You can also recode any variables as planned in your data dictionary (e.g., a reverse coded item) (see Figure 14.9)
13. Construct additional variables.
    - This is not the time to construct analysis-specific variables. This is the time to create or calculate variables that should always be a part of the core study dataset. These variables should be chosen by your data management working group early on and added to your data dictionary. Examples of variables you might add:
        - time components (e.g., `wave`)
        - grouping variables (e.g., `treatment`)
        - foreign keys (e.g., `sch_id` in a teacher file)
        - measure composite or summary scores
        - completion variables or data quality flags

t_stress4		t_stress4_r
1	→	5
2	→	4
5	→	1

FIGURE 14.9
Reverse coding a variable.

- variables created for scoring purposes (e.g., `age`)
- variables that you want added to the core sharing dataset (e.g., categorizing an open-ended text response variable)

NOTE

Some of these variables may exist in other sources (e.g., `treatment` may exist in your participant tracking database). If so, these variables won't need to be created or calculated, they can simply be merged into your clean dataset using a technique similar to the one described in data de-identification step. You can export a file from your participant tracking database that contains a unique identifier (e.g., `tch_id`) as well as the variables you need (e.g., `treatment`), and join that file with your clean data file using the unique identifier.

14. Add missing values.
    - Assign missing value codes based on your designated schema (as documented in your data dictionary and style guide).
15. Add metadata (UK Data Service 2023).
    - While interoperable file types (e.g., CSV) are highly recommended for storing data, it can be extremely helpful to create another copy of your clean data in a format, such as SPSS, that allows for embedded metadata. These file types allow you to embed variable and value code labels that can be very handy for a data user.
    - This can be especially helpful if you plan to export your variables with numeric values (1 | 0), rather than text values ("yes" | "no"). In this case, rather than having to flip back and forth between a file and a data dictionary to interpret codes, users can review information about the variables within the file itself.
    - While future data users may not have a license for the proprietary file type, these file formats can often be opened in free/open source software (e.g., GNU PSPP) or can be easily imported into a variety of other statistical programs that can interpret the metadata (e.g., importing SPSS files into R or Stata).
16. Data validation.
    - It is good practice to assume that some amount of error is inevitable, even with the best data management practices in place. Errors in the data can happen for many reasons, some

of which come from mistakes during the data collection and capture process, others come from the data cleaning process (e.g., coding errors, calculation errors, joining errors). Yet, errors won't be found if you don't actively look for them (Palmer 2023). At a minimum you should always validate, or check, your data for errors at the end of your data cleaning process. Ideally, though, you should be checking every one of your transformations along the way as well. This can be something as simple as checking category counts of categorical variables, or means of numeric variables, both before and after transformations.

- After your data is cleaned, begin with the manual method of opening your data and eyeballing it. Believe it or not, this can actually be a very useful error-catching technique. However, it should not be your only technique. You should also create tables, calculate summary and reliability statistics, and create univariate and bivariate plots to search for errors. Codebooks are great documents for summarizing and reviewing a lot of this information (Arslan 2019).
- You can organize your data validation process by our data quality criteria. The following is a sampling of checks you should complete during your validation process, organized by our criteria (CESSDA Training Team 2017; ICPSR 2020; Strand 2021; Reynolds, Schatschneider, and Logan 2022; UK Data Service 2023).

    1. Complete
        - Check again for missing cases/duplicate cases.
            - It can also be helpful to check Ns by cluster variables for completeness (e.g., number of students per teacher, number of teachers per school) (DeCoster 2023).
        - Check for missing columns/too many columns.
    2. Valid and Consistent
        - Check variables for unallowed categories or values out of range.
            - Checking by groups can also help illuminate issues (e.g., check age by grade level) (Riederer 2021).
        - Check for invalid, non-unique, or missing study IDs.
        - Check for incorrect variable types.
        - Check for incorrect formatting.
        - Check missing values (i.e., do they align with variable universe rules and skip patterns).

3. Accurate
   - Cross check for agreement across variables (e.g., a student in 2nd grade should be associated with a 2nd-grade teacher).
   - Check for other project-specific unique situations.
4. De-identified
   - Are all direct identifiers removed?
5. Interpretable
   - Are all variables correctly named?
   - If applicable, is metadata applied to all variables? Is the metadata accurate (e.g., value labels correct, variable labels correct)?

- If during your validation process you find errors, you first want to determine where the errors originated and correct them in the appropriate location. Errors may come from the following sources:
  1. Documentation errors
     - If there is an error in documentation (e.g., you added a variable to your data dictionary that you did not actually collect, the ranges set for a variable were incorrect), correct the error in your documentation.
  2. Data cleaning errors
     - If an error occurred in your cleaning process (e.g., a variable was recoded incorrectly, a transformation was missed in the cleaning process), fix the error in your process and begin cleaning again at step 1.
  3. Data capture errors
     - If the error occurred in your data entry or data export process (e.g., a variable value was incorrectly entered by a team member, a file was exported using the incorrect export options), fix the error at the source and export a new file. Begin your cleaning process again at step 1.
     - If the error is found in an externally captured source (e.g., student education records), consider reaching out to that provider to determine if there was an error in their process that can potentially be corrected on their end. If a new file is extracted, begin the cleaning process again at step 1.
  4. If you find true values that are inaccurate, uninterpretable, or outside of a valid range (i.e., they represent what the participant actually reported), you will need to make a personal

decision on how to deal with those. No matter what choices your team makes in this process, **make sure they are documented** in the appropriate places for future users (e.g., data dictionary, data cleaning plan, research protocol). Some examples of how you might deal with true errors include:

- Leave the data as is, make a note of the errors in documentation, and allow future researchers to deal with those values during the analysis process.
- Assign a value code (e.g., "inaccurate value" = -90) to recode those values to.
- Create data quality indicator variables to denote which cells have untrustworthy values (e.g., `age` contains the true values and `age_q` contains "no concerns" = 0 | "quality concerns" = 1).
- If you find inconsistencies across different sources, you could choose one form as your source of truth and recode values based on that form.
- If there are true errors where the correct answer can be easily inferred (e.g., a 3-item rank order question is completed as 1, 2, 4), sometimes logical or deductive editing can be used in those cases and the value is replaced with the logical correction (IPUMS USA 2023; Seastrom 2002).

At this point, your dataset should be clean. However, there may be additional transformations to be performed depending on how you plan to store and/or share your datasets.

17. Join data
    - You may have other datasets in your study that need to be combined with your current data. Recall from Section 3.3, that there are two ways you may need to join data, horizontally or vertically.
      1. Joining data horizontally (merging)
         - Merging is commonly used to link longitudinal data within participants in wide format. In this case it will be necessary to append a time component to your time varying variable names if they are not already included (e.g., "w1_", "w2_").
         - Merging can also be used to link forms within time (e.g., student survey and student assessment) or link forms across participant groups (e.g., link student data with teacher data).

- When merging data, it is important that both files contain a linking key (i.e., a unique identifier that will allow you to join files) and that all other variable names are unique across files. Duplicate variable names will not be allowed when merging data.

2. Joining data vertically (appending)
   - Appending may be used to combine longitudinal data within participants in long format. Here it will be necessary to include a new variable that indicates the time period associated with each row.
   - However, appending is also often used for combining comparable forms collected from different links or captured in separate tables (e.g., data collected across sites or cohorts).
   - When appending data, it is imperative that variables are identically named and formatted across files (i.e., same data types, same value codes, same format).

- Depending on how your data is collected or captured, as well as how you want to structure your data, you may use a combination of both merging and appending to create your desired dataset.
- Once your merging or appending is complete, it will be very important to do additional validation checks. Do you have the correct number of rows and columns after merging or appending?

18. Reshape data
    - Recall Section 3.3.2 where we reviewed various reasons for structuring your data in wide or long format.
      1. In wide format, all data collected on a unique subject will be in one row. Here, unique identifiers should not repeat.
      2. In long format, participant identifiers can repeat, and unique rows are identified through a combination of variables (e.g., `stu_id` and `wave` together).
    - If at some point after merging or appending your data, you find you need to reshape data into a new format, this restructuring process will need to be added to your data cleaning process.

NOTE

If working with longitudinal data, concatenating the time component to the beginning or end of a variable name (as it is in Figure 14.10), rather than embedding it into your variable name, makes this back and forth restructuring process much easier to do in statistical programs.

Wide format

tch_id	w1_t_stress1	w2_t_stress1
236	1	2
238	4	3

Long format

tch_id	wave	t_stress1
236	1	1
238	1	4
236	2	2
238	2	3

FIGURE 14.10
A comparison of long and wide format.

19. Save your clean data
    - The final step of your cleaning process will be to export or save your clean data. You can save your files in one or more file types depending on your needs. It can be helpful to save your data in more than one format to meet various analysis, long-term storage, or data sharing needs (e.g., an interoperable format like CSV, and a format that contains embedded metadata such as SPSS).

14.4 Data Cleaning Workflow

Data cleaning is not a standalone process. It should be part of a larger, well-planned workflow that is designed to produce standardized, reproducible, and reliable datasets. Ignoring this planning and jumping into data cleaning in a haphazard way only leads to more work after the cleaning process, having to organize messy work so that others can understand what was done. Assign someone to oversee and implement this workflow and designate a time frame for when data cleaning should both begin and be finalized each wave. Then, document this information in the SOP associated with each data source. A high-level summary of your data cleaning procedures should also exist in documents such as your data management plan and your research protocol.

14.4.1 Preliminary Steps

The first part in creating a data cleaning workflow is making sure that your folder structure is set up according to your style guide, and that your folders and files are consistently named according to your style guide. It is also important that the metadata in your names is always provided in the same order (e.g., project -> time -> participant -> instrument -> type). Breaking away from a standardized naming convention begins to erode the reproducibility of your work.

Next, you will want to gather all of the necessary documentation that will be used throughout your cleaning process.

1. Data dictionary
   - In this document, variables should be named and coded according to your style guide and all variables and transformations have been approved by the data management working group.
2. Data cleaning plan
   - This should include a series of steps based off of our standardized data cleaning checklist, and all transformations have been reviewed by the data management working group.
3. README files
   - This includes any README files, stored alongside raw data files, that contain notes that may be relevant to your data cleaning process (e.g., a project coordinator notes that "ID 1234 should actually be ID 1235"). You will want to integrate this information into your data cleaning plan as needed.
4. Participant tracking database
   - Make sure that this database is up to date so that you can compare form completion status numbers to the Ns in your dataset.

Once you gather your documentation, you are ready to begin the data cleaning process.

14.4.2 Cleaning Data Using Code

While you can clean data through a point and click method in a program like SPSS or Microsoft Excel, cleaning data manually is typically not reproducible, leads to errors, and is time consuming. The number one practice that you can implement to improve the reproducibility, reliability, and efficiency of this workflow is to clean data using code (Borer et al. 2009). The code can be written in any program your team chooses (e.g., R, SAS, Stata, SPSS) and saved in a syntax, or script file, that can then be re-run again at any point. While writing code may seem time consuming up front, it has numerous benefits.

- It can actually save you an enormous amount of time in the future if you plan to clean data for the same form multiple times (in say a longitudinal study).

- Being able to see and track each of your steps helps you to be more thoughtful in your data cleaning process.
- It allows others to review your process in order to provide feedback and catch potential errors.
- It makes your work more easily reproducible. By simply re-running your code file, anyone should be able to get the same resulting dataset that you created.

However, writing code alone will not provide all of the desired benefits. There is more that must be considered.

1. Choose an appropriate tool to code in. Assess things such as:
   - Your comfort level with the program as well as available support.
   - Cost and access to the program.
   - Interoperability of the program (i.e., will others be able to open, review, and run your code).
   - Limitations (e.g., file size limitations, variable character count limitations).
   - Default settings (e.g., how the program performs rounding, how dates are stored).
2. Follow a coding style guide.
   - As discussed in Section 9.6, coding best practices such as using relative file paths, including comments, and recording session information, reduces errors and allows your processes to be more reproducible. Adding best practices to a code style guide ensures that all team members are setting up their files in a consistent manner, further improving the usability of code.
3. Review your data upon import.
   - As we discussed in Section 14.3.1, it is imperative that you review your data before beginning to clean it to ensure you have a thorough understanding of what is happening in your file. This review process can become even more relevant if you are reusing a syntax file to clean data collected multiple times (e.g., in a longitudinal study). You may expect your syntax to run flawlessly each time period, yet if anything changes in the data collection or entry process (e.g., a variable name changed, a new item is added, a new variable category is added), your data cleaning syntax will no longer work as intended. It's best to find this out before you start the cleaning process so you can adjust your data cleaning plan and your code as needed.

4. Do all transformations in code.
   - Cleaning data using code only improves reproducibility if you do all transformations, no matter how small, in the code. No transformations should be done to your data outside of code, even if you think it is something insignificant. Once you work outside of your code, your chain of processing is lost, and your work is no longer reproducible. Code files should contain every transformation you make from the raw data to your clean data.
5. Don't do anything random.
   - Everything in your syntax must be replicable. Yet, there are a few scenarios where, without even realizing it, you could be producing different results each time you run your code.
     - If you randomly generate any numbers in your data (e.g., study IDs), use an algorithmic pseudorandom number generator (PRNG). This can be easily done in most statistical programs by setting a seed. Every time the PRNG is run with the same seed, it will produce the same results (i.e., the same set of random numbers). Without this, you will get a new random set of numbers each time your syntax is run.
     - Another example is when you are removing duplicate cases. Be purposeful about how you remove those duplicates. Do not assume your raw data will always come in the same order. Set parameters in your syntax before dropping cases (e.g., order by date then drop the second occurrence of a case). Otherwise, if at some point, someone unexpectedly shuffles your raw data around and you re-run your syntax, you may end up dropping different duplicate cases.
6. Check each transformation.
   - As mentioned in Section 14.3.1, check your work along the way; don't wait until the end of your script. For each transformation in your data:
     - Review your variables/cases before and after the transformations. Did the transformation work as you expected it to?
     - Review all errors and warning codes. Some messages will be innocuous, while others may be telling you that something unexpected happened.
7. Validate your data before exporting and review after exporting.
   - As we discussed in Section 14.3.1, before exporting data you will want to run through your final list of sanity checks, based on our data quality criteria, to make sure no mistakes are missed.

    - While eyeballing summary information is helpful, consider writing tests based on your expectations, that produce a result of TRUE or FALSE (e.g., test that `stu_id` falls within the range of *1000–2000*).
  - After exporting your data, open the exported file. Does everything look as you expected (e.g., maybe you expected missing data to export as blanks, but they exported as "NA")?

8. Do code review.
   - If you have more than one person on your team who understands code, code review is a great practice to integrate into your workflow. This is the process of having someone, other than yourself, review your code for things such as readability, usability, and efficiency. Through code review it's possible to create more interpretable code as well as catch errors you were not aware of. Code review checklists can be implemented to standardize this process.

Resources

Source	Resource
Travis Gerke	R Code Review Checklist[1]

14.4.3 Cleaning Data Manually

While cleaning data with code is the preferred method for the reasons previously mentioned, it does require technical expertise that not every team may have. If your team needs to clean data manually, consider two important things.

1. Choose a tool based on the same criteria used when choosing a coding tool (i.e., comfort level, cost and access, interoperability, and default settings). Be aware of the potential formatting issues mentioned in Section 12.2 when cleaning with tools like Microsoft Excel.
2. Once you begin cleaning your data manually, it is imperative that you document every transformation to enable reproducibility. This may look different depending on the tool you use.
   - If cleaning data using the point and click menu in a program such as SPSS, when performing a transformation use a "paste" type button to copy all associated commands into a syntax file that can easily be reused (Kathawalla, Silverstein, and Syed 2021).

- If using a program such as Microsoft Excel for data cleaning, add notes into your data cleaning plan that are detailed enough to allow anyone to replicate your exact data cleaning process by hand (Hoyt et al. 2023).

14.4.4 Data Versioning Practices

The last part of the workflow to consider is where you will store your data and how you will version it. First, as you export or save your clean datasets, make sure to name them appropriately to differentiate between raw and clean datasets. During an active project, it is typically best to store clean datasets in their respective individual folders (e.g., wave1 -> student -> survey -> clean folder), rather than moving all clean files to a separate "master folder". If changes need to be made to files, it is easier to keep track of files in their original locations. Ultimately, though, what is most important is to not copy files across folders during an active project. Keep one single copy of each clean dataset for authenticity purposes (CESSDA Training Team 2017; UK Data Service 2023). At the end of your project, you may consider moving or copying files to a "master folder" for archiving purposes (see Section 15.2).

After your clean datasets are saved, it is common to find an error in your data and/or your code at a later point. Yet, once you've begun using your data or you've shared it with others, it will be imperative that you do not save over existing versions of those files. You will need to version both your code and your data, following the guidelines laid out in your style guide. Versioning your file names, and keeping track of those different versions in a changelog (see Section 8.3.2), allows you to track data lineage, helping users understand where the data originated as well as all transformations made to the data. While you can version any files that you choose, I am specifically referring to final files here, not in-progress, working files that have not yet been used or shared with others.

Last, along with assigning someone to oversee data cleaning, it will be important to assign someone to oversee this versioning process. Versioning files and updating documentation takes time and consideration, and that responsibility will need to be explicitly laid out in order to ensure it isn't forgotten.

Note

1 https://github.com/tgerke/r-code-review-checklist

15

Data Archiving

Once you have gone through all cycles of data collection and you are preparing to wrap up your grant, you will need to switch gears and start thinking about archiving your data to ensure that your files are still accessible and usable, long after a project is complete (see Figure 15.1). While your project may be ending, there are still many reasons to retain your data long-term, including future analyses, opportunities to make corrections (e.g., going back to paper files if an error is found in the data), and data retention requirements from funders and institutions. In this section we will discuss how to care for files internally, while public data sharing and archiving will be discussed in Chapter 16. However, these processes (preparing to internally archive and preparing to publicly share data) will most likely be happening at the same time.

15.1 Long-Term Storage

The first thing to do when planning to store your data long-term is review your requirements. There may be requirements for both data retention and destruction depending on your oversight (see Section 4.3). It is common for oversight to require that you retain your data anywhere from three to ten years, and there may be specific destruction requirements for data that contain personally identifiable information (PII) or data covered by a data use agreement. Make sure to review relevant policies and agreements to determine what is required.

Once you understand requirements, make a plan for retention and destruction. If you are required to retain your data for a specified number of years, consider how you will continue to store data and documents in a way that meets your original goals (e.g., data safety, protecting confidentiality, accessibility, and usability of files). You will need to consider both paper and electronic files.

15.1.1 Paper Data

At the end of your project, you will want to begin boxing up paper files for long-term storage. Make sure to clearly label all boxes in case you need to

DOI: 10.1201/9781032622835-15

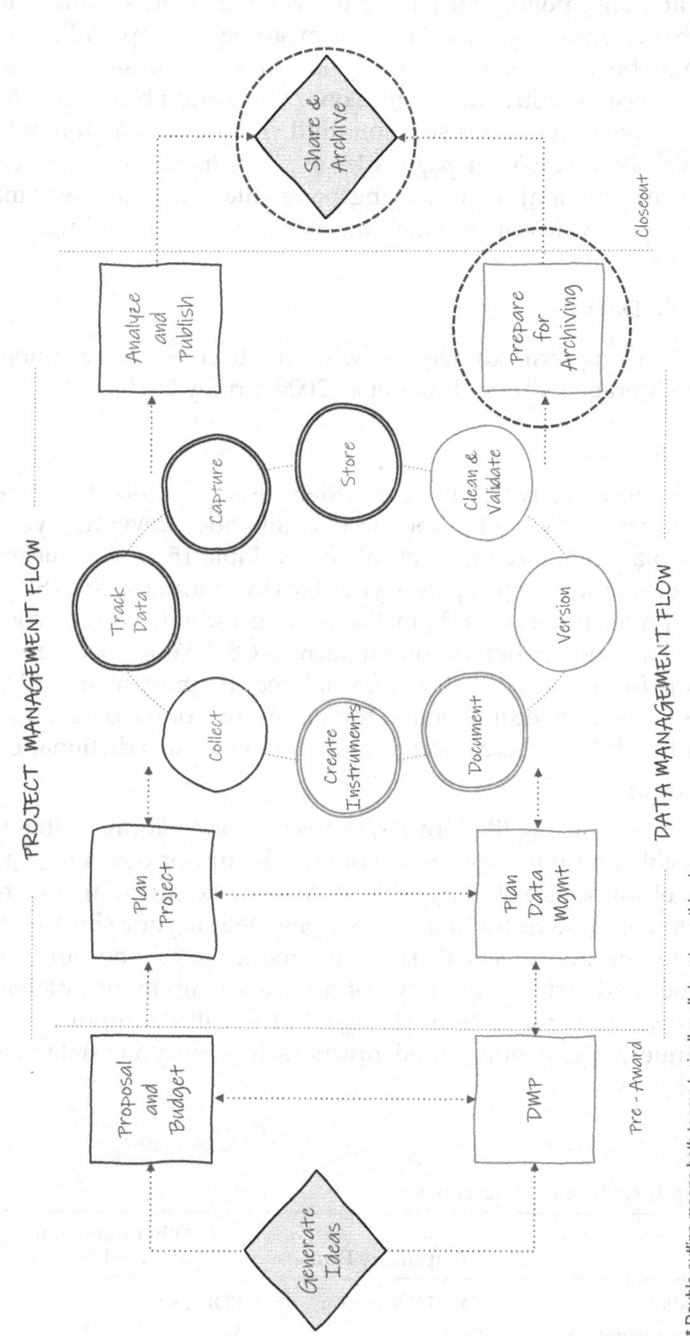

FIGURE 15.1
Long-term data storage in the research project life cycle.

return to files at a later point. Many institutions have records management departments that can assist you with long-term storage of paper files. These departments may be able to store your physical files for a designated set of time, as well as assist in destruction of files once that period has ended. Note that you may not want to use this solution until you are certain you will no longer need to easily access your paper files (e.g., for fixing errors, entering any additional information). If destroying paper files on your own, make sure to choose a quality destruction method such as paper shredding.

15.1.2 Electronic Data

For electronic data long-term storage, you will want to consider two things—file formats and storage location (Borer et al. 2009; Briney 2015).

- File formats
  - First, choose file types that are widely used (i.e., don't require proprietary software) for both accessibility and preventing your file formats from becoming obsolete (see Table 15.1). This means that you can still keep copies of your files in a format such as SPSS if you prefer, but it is good practice to have a second copy of your data in a non-proprietary format such as CSV. Your documentation file formats should also be considered. Formats such as PDF or TXT are often recommended for long-term storage of text documents, while CSV is a good format for tabular data dictionaries.
- Storage location
  - Similar to choosing file formats, choose a storage location that is accessible and not at risk of becoming corrupt or obsolete (e.g., think obsolescence of floppy disks). Also, make sure that you are able to continue restricting access as needed. If your short-term storage solution meets these requirements (e.g., your institution network drive), you may not need to do anything different in preparing for long-term storage, but it will be important to continue implementing good practices to keep your data safe

TABLE 15.1

Potential Long-Term Storage File Types

Type	Non-Proprietary Formats	Other Commonly Accepted Formats
Text documentation	TXT, HTML, XML	PDF, DOCX
Rectangular documentation	CSV	XLSX
Datasets	CSV, TSV	SPSS, STATA, SAS, R, XLSX

(e.g., continuing data backups, checking that hardware and software are up to date).

- Within your storage location, consider copying all finalized datasets (i.e., cleaned and de-identified) into a "master data" folder for ease of future accessibility. Restrict access to reduce unintended modification of files.
  - Design this "master data" folder like you would a public repository folder (see Section 16.3).
    - Add a README that describes what files the folder contains.
    - Copy relevant documentation from other locations to this folder (e.g., data dictionaries, project-level documentation).

When it comes time to destroy data, make sure to permanently delete files, including all backups of files. When deleting PII, this often involves more than just moving files to the recycle bin on your computer. Work with your institution IT department during this process.

15.1.3 Oversight and Documentation

Last, make sure to document your plan, including time frames for retention and destruction, in the appropriate locations (e.g., DMP, research protocol, informed consent agreements, team data security policy). Assign and document responsibilities for short-term tasks such as boxing or relocation of files, as well as ongoing long-term tasks such as maintenance or destruction of files.

Resources

Source	Resource
Kristin Briney	Project Close-Out Checklist for Research Data[1]

15.2 Internal Data Use

At the end of a project, and often even earlier in a project, team members will want to begin analyzing data. It is important to consider how you will make team members aware of what data is available and how you will allow team members, and other research collaborators, to access data. Most likely you will not want researchers going into folders and grabbing datasets without consulting with a core team member first. Therefore, it is important to develop a system for providing data to researchers on an as-needed basis. Create a data request process for team members, or external collaborators, to request access to finalized study datasets. In developing this system, several things will need to be considered.

1. Add descriptions of the finalized datasets to a data inventory to inform your team of their availability (see Section 8.1.4). If possible, link to data dictionaries and other documentation to allow researchers to review information before submitting requests.
2. Design a system for requesting access (e.g., designate a person to email, develop a form that is submitted to a designated person).
    - In this system, the researcher should be asked to describe what data they are requesting, including the exact variables they need and from what time periods, as well as the purpose of their analysis.
    - This system should also include collecting any required agreement forms from requestors (e.g., data sharing agreements).
3. Decide who needs to review the request to ensure the application is complete (e.g., a data manager), and who needs to give final approval for the data request submission (e.g., a principal investigator).
4. Design a system for gathering data for requestors (e.g., provide researchers with full datasets, narrow datasets based on specific requests).
    - If narrowing datasets for researchers, where will new datasets be stored? One option is to create a "data request" folder where you store all data request datasets and accompanying documentation.
5. Consider how you will share datasets with researchers (e.g., a secure link to a cloud folder, using secure file transfer).
6. Consider how you will track data requests.
    - It is important to keep track of data requests in case you need to reach back out to researchers for situations such as errors found in the data. One way to do this is to keep a data request log. The log can include information such as name of researcher, their institution, their email, date of request, name of the project they are requesting data for, and any other helpful tracking information (e.g., data sharing agreement received, request approved, dataset shared with researcher). Using this log, you can reach back out to researchers as needed to inform them of any updates regarding the shared data.

Here is an example of how you might structure a data request folder.

```
data_requests/
├── data_request_log.xlsx
├── lastname1_firstname1
│   ├── projname_data-sharing-agreement_lname1-
│       fname1_2023-04-08.pdf
│   ├── projname_stu_svy_clean_lname1-fname1_v01.csv
│   └── projname_stu_svy_data-dictionary.xlsx
├── lastname2_firstname2
│   ├── archive
│   │   ├── changelog.xlsx
│   │   └── projname_tch_svy_clean_lname2-fname2_v01.csv
│   ├── projname_data-sharing-agreement_lname2-
│       fname2_2023-04-22.pdf
│   ├── projname_tch_svy_clean_lname2-fname2_v02.csv
│   └── projname_tch_svy_data-dictionary.xlsx
└── ...
```

Ultimately, you want to establish a standardized and efficient process that reduces the burden on team members responsible for reviewing, approving, and fulfilling data requests and also removes any ambiguity about how users should request access to data (Institute of Education Sciences 2019). As always, roles and responsibilities will need to be assigned to each step of this process. Often the person who facilitates internal data requests for one project will likely be the same person who fulfills data requests for all projects. Document this request process in your data security policy (see Section 8.1.5) so that team members know how to request data and who to work with.

Resources

Source	Resource
Crystal Lewis	Sample data sharing agreement and request form[2]

15.3 Using a Repository

Last, if maintaining your electronic data long-term sounds like too much effort for your team, there are other options. Many universities have institutional repositories that may include services such as data curation and preservation. Additionally, there are several external repositories that offer curation and preservation services where you may be able to deposit your data for long-term storage. It's possible that archiving your data using one of these two options may also align with publicly sharing your data, which we will review in Chapter 16.

Notes

1 https://authors.library.caltech.edu/records/yr0y9-z4q70
2 https://osf.io/ybdw6

16

Data Sharing

Throughout a project, teams are internally sharing data and materials with a variety of people (e.g., team members, collaborators, funders) who use that information for a variety of purposes (e.g., analyses, reports, to answer questions). Yet, at the end of a project, or possibly earlier, it's important for researchers to also consider making their research data available for broader public use (see Figure 16.1). However, publicly sharing project data and materials requires a lot of consideration. In this chapter we will first review reasons why you should publicly share your data, and then we will work through a series of decisions to make before sharing your data.

16.1 Why Share Your Data?

The most notable reason for openly sharing data is that there are a growing number of supporting organizations (e.g., funders, journals, institutions) that are requiring researchers to share data. Federal agencies in particular want to ensure that there is free, open access to taxpayer-funded funded research (Nelson 2022). Beyond requirements though, there are many other reasons researchers should want to share their data. One, it benefits the scientific community by improving rigor. Sharing both data and code helps to discourage fabrication and encourages validation of results through both replication and reproducibility of findings (Alston and Rick 2021; Cook et al. 2021; Gonzales, Carson, and Holmes 2022; Institute of Education Sciences 2023b; Klein et al. 2018; Levenstein and Lyle 2018; Logan, Hart, and Schatschneider 2021; Meyer 2018). Allowing other researchers to reuse your data also reduces the need to duplicate data collection efforts, saving time, energy, and money (Gonzales, Carson, and Holmes 2022; ICPSR 2020; Levenstein and Lyle 2018), as well as reducing the burden on the communities that are frequently targeted for data collection (Gaddy and Scott 2020). It also encourages diversity of analysis and perspectives (ICPSR 2020; Levenstein and Lyle 2018). Not only may researchers have novel questions not considered by the original

DOI: 10.1201/9781032622835-16

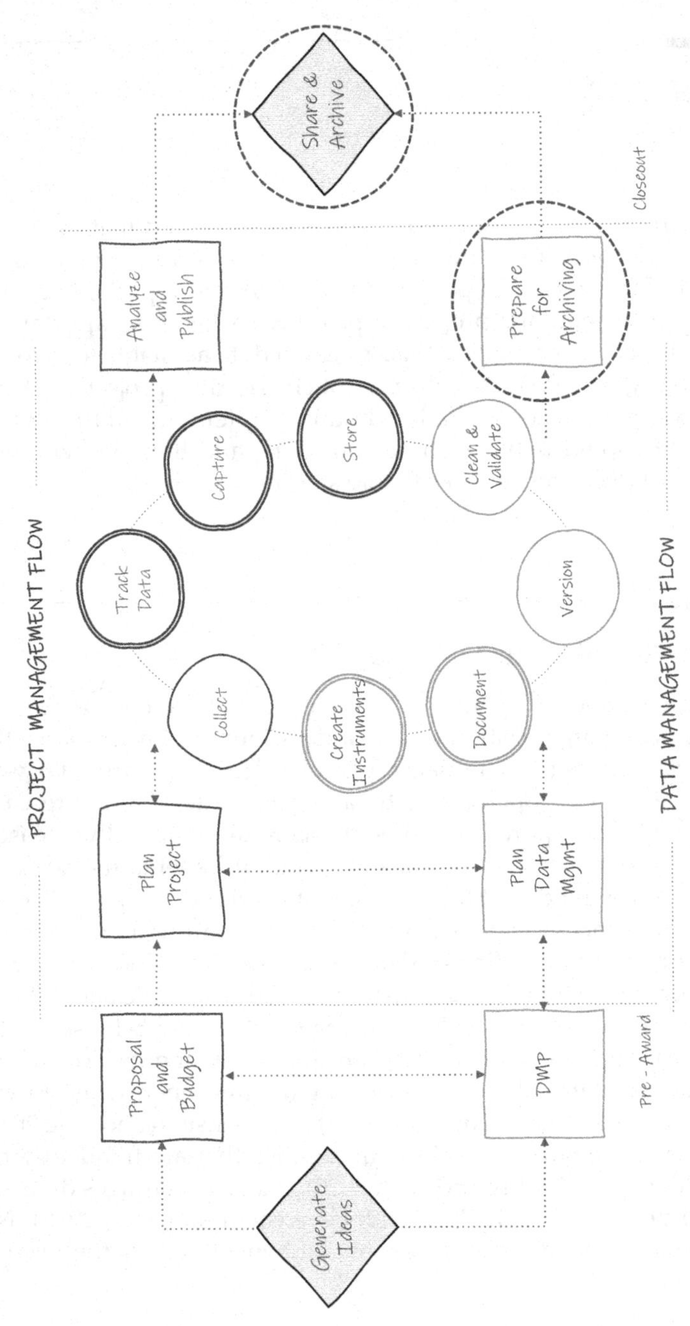

* Double outline means both teams typically collaborate on this phase

FIGURE 16.1
Data sharing in the research project life cycle.

investigators, open data also provides the opportunity for researchers to improve upon or experiment with new methods, as well as combine datasets to facilitate new discoveries (Cook et al. 2021; Institute of Education Sciences 2023b; Logan, Hart, and Schatschneider 2021; Meyer 2018). Openly sharing data also provides equitable access to high-quality datasets for early career scholars, students, and underrepresented researchers who otherwise may not have the budget, staff, or connections to collect data (ICPSR 2020; Logan, Hart, and Schatschneider 2021). Last, data sharing can have the unintended benefit of promoting more efficient and sustainable data management practices (Klein et al. 2018). Knowing that data and documentation will eventually be shared outside of the team may motivate researchers to think hard about how to organize their data management practices in a way that will produce materials that they are comfortable sharing publicly. This is not to say that data sharing is without challenges, including the additional time, energy, and resources required by a team to prepare data for sharing. However, planning your data management practices around your data sharing plan early on can help reduce any significant burden that may otherwise be caused by data sharing (Klein et al. 2018; Levenstein and Lyle 2018).

16.2 Data Sharing Flow Chart

There is a series of decisions to be made when sharing your data (see Figure 16.2). While in some cases data sharing may not occur until the end of your project, many of your data sharing decisions will actually need to be made at the beginning of your project, when you write your data management plan (DMP) (see Chapter 5). These decisions will inform both your workflow during your active project and the steps you need to perform when preparing your data for archiving.

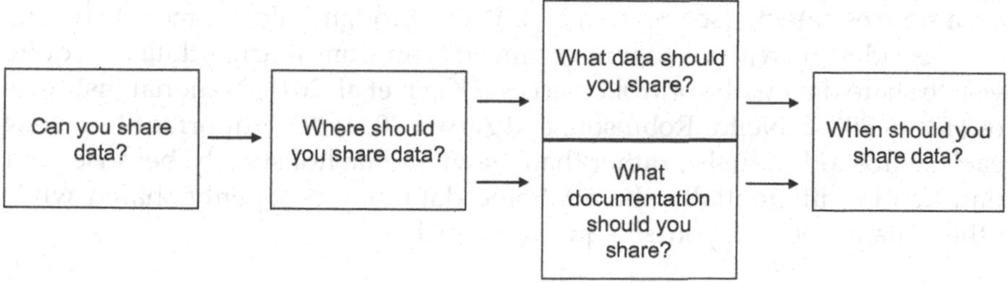

FIGURE 16.2
Decisions to be made before publicly sharing study data.

In this section we will walk through the flow chart, discussing what information is needed to make decisions as well as best practices associated with each decision.

16.2.1 Are You Able to Share?

When it comes to data, there are three degrees of sharing (Ghent University 2023).

1. Open data
    - This is data that can be publicly shared with no constraints. These are typically data with no ownership concerns and that have minimal disclosure risk (e.g., de-identified, not highly sensitive).
2. Controlled data
    - These are data that are not openly shared, but can be shared in other ways under restricted access and use conditions. This typically involves a data request system including an application and/or data use agreements, and only after a request is approved can data be accessed. This includes data with more than minimal disclosure risk (e.g., highly sensitive information, identifiable information, information that data owners want shared in more restrictive ways).
3. Closed data
    - These are data that cannot be shared due to legal, ethical, or technical reasons (e.g., proprietary data, data use agreements forbid it, participant consent does not allow data sharing).
    - In these cases, it may still be possible to share some data (i.e., summary statistics, metadata, documentation), to allow the information to be discoverable, reusable, and citable (Logan, Hart, and Schatschneider 2021; Neild, Robinson, and Agufa 2022).

When preparing to make data sharing decisions, it is helpful to pull out your data sources catalog (see Section 5.3). Walk through each source one by one and consider the reasons that may prevent you from sharing data or require you to share data with controlled access (Klein et al. 2018; National Institutes of Health 2023c; Neild, Robinson, and Agufa 2022). It's important to review each source individually, rather than viewing data holistically, because your data can be differentially shared. Some data may be openly shared while other data is not. Ask yourself questions such as:

- Who is the owner of this data? Do I have permission to share this data?

    - This may involve reviewing data sharing agreements, licenses, or other documents.
- Do I need consent to share this data? If yes, does my consent form include data sharing language (see Section 11.2.5)?
- Will I be able to de-identify data to the point where there is minimal disclosure risk?
- Are there other compelling legal, ethical, or technical reasons to not share your study data?

It is important to make decisions based on the philosophy of "as open as possible, as closed as necessary" (European Commission. Directorate-General for Research & Innovation 2016, 4). We want data to be open to facilitate reuse and garner all of the benefits mentioned in Section 16.1, but data should be closed when necessary to protect the privacy of individuals or to honor other prior agreements. However, many of the reasons for not being able to share your data at all can be mitigated by early planning. Developing consents to clearly explain data sharing plans, and talking with partners early on about plans to share data, can help increase the amount of data you can openly share (Klein et al. 2018; Neild, Robinson, and Agufa 2022).

16.2.2 Where to Share?

Once you've decided that you are able to share data, either publicly or in a more restrictive manner, you then need to decide where you want to share data. There are many options for sharing your data, with some options being better than others. In most situations, the best option for sharing data will be a public data repository. This may be a repository chosen by you, or it may be a repository designated by a funder or other organization supporting your work. There are many benefits to sharing in a public repository (Neild, Robinson, and Agufa 2022; UK Data Service 2023).

- It meets the FAIR principles of making your data both findable and accessible (Section 2.4.1).
- It is the preferred method of data sharing for many supporters (e.g., NIH, IES) and it may be required by others (e.g., NIJ, NIMH, journals such as *AMPPS*).
- It provides a hands-off approach to data sharing, reducing the burden of maintaining your data and responding to data requests long-term.
- It provides a means to securely share restricted-access data if necessary.

- Even if you are not able to directly deposit the restricted-use data in a repository (e.g., the repository does not accept restricted-use data, an agency partner requires data to be stored at their site), repositories still support the creation of metadata-only records, which facilitate discovery of your data while still allowing sensitive data to be maintained and shared through the owner's chosen data request system (Gonzales, Carson, and Holmes 2022; Logan, Hart, and Schatschneider 2021).
- Repositories provide support for data sharing, either through direct data curation services or by offering detailed guidelines on what to share.

However, due to supporter requirements (e.g., agency partner, institution, funder, journal), or other legal, technical, or ethical reasons, there may be a reason to share some or all of your data in other ways. The following are alternative ways to share your data (Alston and Rick 2021; Briney 2015; Klein et al. 2018; Neild, Robinson, and Agufa 2022; UK Data Service 2023).

1. Deposit your data with an institutional archive.
    - While this method provides the benefits of reducing the burden on your staff and securely storing your data, this option is not available at all academic institutions and these repositories may provide less service offerings than a public repository. Furthermore, data stored in an institutional archive, as opposed to a public repository, may be less findable for researchers outside of your institution.
2. Deposit with a partner agency.
    - In some cases, data sharing may only be allowed if you deposit the data with an agency that you partnered with for the study (e.g., a school district). In this case, all data requests will go through that partner.
3. Supplemental materials attached to an article or stored on a publisher's website.
    - A concern with this method is that materials will be lost if a journal changes publishers or a publisher changes its website.
4. Share through a lab, personal, or project website.
    - Using this method you may allow participants to freely download datasets from the site, or they may be required to go through an application process first before accessing data.
    - While this system provides some accessibility, it also requires a significant commitment from your team. Sharing in this way

requires to you publicize your site to increase visibility, as well as commit the resources to building and maintaining a secure and reliable data sharing pipeline. Furthermore, websites change and links break, reducing the long-term accessibility of your data.

5. Informal peer-to-peer sharing.
    - Here data owners share data with peers on an as-needed basis. While this method may work fine for peers, it doesn't make a broader audience aware of the availability of your data and keeps the burden of data maintenance and responding to requests on you and your staff.
    - This type of sharing is often synonymous with using a "data available upon request" statement in a publication. While these statements do make broader audiences aware of your data, there are several studies that have found that these data availability statements rarely result in access to data (Stodden, Seiler, and Ma 2018; Vines et al. 2014). A better alternative is to already have data shared in a location and direct people to that system (e.g., link to a repository).

Figure 16.3, modified from a flow chart created by Borghi and Van Gulick (2022), can help you work through the process of choosing where to share each data source. Ultimately though, no matter where you choose to share your data, it is important to make this decision early on because it will impact many of the other decisions that need to be made. In particular, if you choose to deposit data in a repository, you will want to review repository-specific requirements and standards to make sure they are accounted for in your DMP (e.g., data format requirements, metadata standards used) and in your data management processes throughout the study. Making this decision early on also allows you to begin creating a schedule for ongoing data deposits throughout your study if that is something you want to consider doing, or if it is required by your funder or other supporter (ICPSR 2020).

16.2.2.1 Choosing a Repository

At this point, you may be ready to choose a repository to share your data in. There is an abundance of available data repositories to choose from and the Registry of Research Data Repositories (re3data.org) has indexed repositories, allowing researchers to search the vast landscape of options. Several agencies have also shared criteria to help you narrow your choices. Both the National Institutes of Health (2023b) and the National Science and Technology Council (2022) have released desirable characteristics for data repositories, and the Institute of Education Sciences has also provided its own set of dimensions

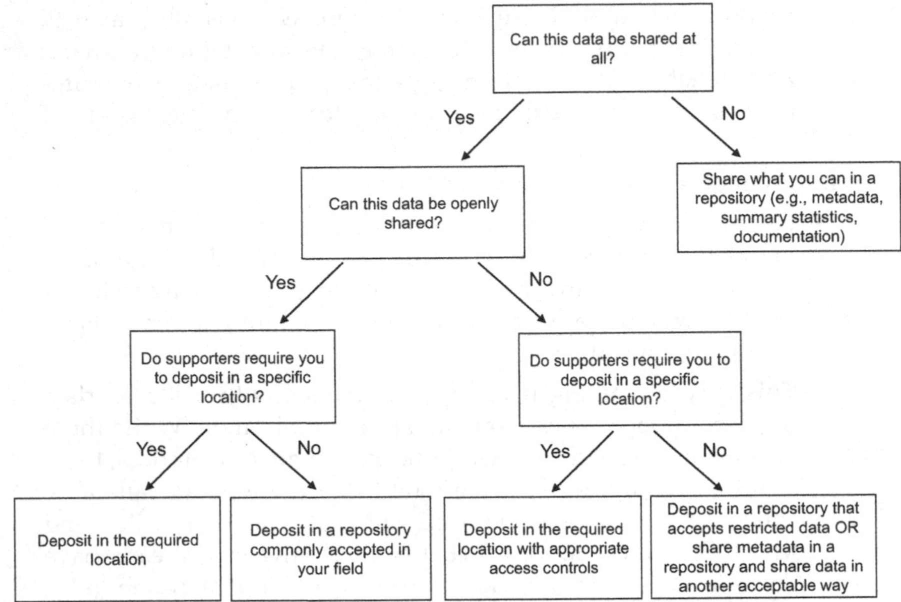

FIGURE 16.3
A series of decisions to make when deciding where to share data.

to review when considering an appropriate repository (Neild, Robinson, and Agufa 2022). While each of these agency lists should be reviewed, the following questions are a starting point for choosing a repository that fits the needs of your project (Briney 2015; Gonzales, Carson, and Holmes 2022; Goodman et al. 2014; Klein et al. 2018).

1. Is a specific repository required by your supporter (e.g., funder, journal, institution)?
    - If yes, you don't need to proceed any further, this is the repository you should share your data in. If the required repository does not meet your specific needs (e.g., you want to reach a specific audience), you can always share metadata in another repository and link to the repository where your data is shared.
    - If a specific repository is not required, also check to see if your supporter has a preferred list of repositories. If they do, it may be best to narrow your search to the recommended options.
2. Is a domain-specific repository available (i.e., caters to your field or specific data type) or are there generalist repositories that are commonly accepted in your field?

- Domain-specific repositories may be of more interest to researchers in your field and may be the best option to help facilitate discovery of your datasets. Using a domain-specific repository can also help ensure that you are preserving data according to recognized standards in your field. At minimum though, using generalist repositories that are common in your field improves discoverability.
- A few domain-specific repositories commonly used in the field of education research include ICPSR (https://www.icpsr.umich.edu) and LDbase (https://www.ldbase.org/).
- NIH has also established the Generalist Repository Ecosystem Initiative (GREI), which consists of established generalist repositories that support FAIR principles and are collaborating to develop a standard set of services and structures. Generalist repositories in this ecosystem that are commonly used in education research include OSF (https://osf.io), Zenodo (https://zenodo.org/), and Figshare (https://figshare.com/).
- If you have qualitative data, there are also repositories specifically designed for this type of data, such as the Qualitative Data Repository (https://qdr.syr.edu/).

3. Does the repository allow varying access levels?
   - Can de-identified data be easily accessed by a wide audience?
     - What are the access options (e.g., download)?
     - Are users required to have an account to access data?
   - Are restricted-use files accepted?
     - Is there a transparent process for reviewing data access requests? Who reviews access requests?
     - How are users able to access restricted-use data files (e.g., secure download, virtual data enclave, physical data enclave)?
4. Is there a cost associated with using the repository?
   - Consider both costs to store your data and costs for users to access your data. There may also be costs associated with additional services such as data curation.
5. What are the allowable file formats and sizes?
   - Check size limits for both data files and the entire project. Also check which file formats are allowed. Certain repositories may have file format preferences for both data files and documentation files.
   - You'll also want to make sure that the repository provides files back to users in commonly accepted formats, including at least one non-proprietary format.

6. Does the repository have long-term sustainability?
   - Make sure to review the repository's data retention policies to ensure it meets your requirements.
   - You'll also want to ensure that the repository has a plan for long-term management of data (considering both funding and infrastructure).
7. Does the repository have linking capabilities?
   - Does the repository allow you to link to projects, publications, code, or data stored on external sites?
8. Are clear use guidelines provided?
   - Is there clear guidance on how data can be used? Does the repository allow you to add usage licenses?
     - Licenses set clear terms of use for data. Commonly used license groups include Creative Commons licenses (https://creativecommons.org/) and Open Data Commons licenses (https://opendatacommons.org/). The most commonly used licenses are CC BY and ODC-By which allow others to freely reuse materials as long as they cite the original creator.
   - Are you able to set different reuse conditions for different datasets?
9. Is metadata collected upon deposit?
   - Does the repository collect comprehensive metadata, and does it use standards that are appropriate for your field?
10. Does the repository assign unique persistent identifiers?
    - Unique persistent identifiers (PIDs), such as digital object identifiers (DOIs), provide an enduring reference to a digital object, even if the object's URL changes. Repositories that assign PIDs support discoverability as well as allow researchers to track the use and contributions of their datasets.
11. Does the repository track data provenance?
    - It is important to have the ability to freely update and remove data as needed.
    - When data is amended, does the repository record data provenance by versioning materials and assigning updated PIDs?
12. Are curation and quality assurance services available?
    - Does the repository provide curation services to ensure that data is de-identified, high quality, in interoperable formats, and shared with appropriate metadata?

13. Does the repository measure reuse?
    - This includes things like tracking number of downloads and tracking citations.
14. Does the repository have appropriate security measures in place?
    - This includes measures for both the security of the data itself (e.g., maintaining backups) and measures to ensure participant privacy (e.g., ensuring only authorized users are able to access restricted data).

Source	Resource
Institute of Education Sciences	Data repository comparison chart on page 27[1]
Shelley Stall, et al.	Generalist repository comparison chart[2]

16.2.3 What Data to Share

The requirements for what data should be shared will most likely vary depending on your supporting agency. As an example, NIH asks researchers to share their final research data which includes any "recorded factual material commonly accepted in the scientific community as necessary to validate and replicate research findings, regardless of whether the data are used to support scholarly publications" (National Institutes of Health 2023c).

Even if your supporter does not require this broad sharing, in general this is still a good policy to follow. At the end of your study, share all of the data collected and captured for the project, while minding any legal, ethical, and technical reasons that you cannot share some information. With the exception of identifying information, this includes sharing all primary data collected through the project, both raw (item-level) and derived variables, as well as any secondary data captured and linked to your sources (e.g., student education records) (ICPSR 2020).

Before sharing, review any existing agreements or licenses associated with each data source to ensure you are allowed to share and in what format you are allowed to share. Also review any applicable consent agreements. You will not want to share any data outside of the scope of what participants agreed to. Last, before sharing item-level data, review copyright for published scales to see what is allowed. Some publishers may only allow you to share derived scores (Logan 2021).

16.2.3.1 Processing of Files

As discussed in Chapter 14, there are three levels of data files, raw, clean, and analytic. When publicly sharing project data, it's often best to not share your raw datasets. This may seem counterintuitive to ideas of transparency and reproducibility, but in education research these raw datasets often contain identifiable information, and despite our best efforts to collect comprehensible data, they typically do not meet our data quality criteria. They tend to still require some sample cleaning (e.g., removing duplicates), as well as some variable renaming, recoding, or other transformations to ensure that data are not misused or misinterpreted by future researchers.

Instead, it is more useful to share the general clean datasets discussed in Chapter 14. These datasets have all direct identifiers removed (indirect identifiers discussed in Section 16.2.3.4), they have been curated to allow for easier interpretation of variables, and they contain all of the information necessary to validate any research findings. In the same repository, it is also possible to share any analytic datasets created that will allow replication of findings in any specific reports or publications.

16.2.3.2 Organizing Files

If you are collecting data across time, across different forms, or across different cohorts, you will want to consider whether or not you want to combine files before sharing or if you want to provide distinct files that users can merge on their own (Neild, Robinson, and Agufa 2022).

- Combined files
  - Combining datasets may be a great option for longitudinal studies with many waves of data collection. Combining files across time can reduce the burden for future researchers who want to view all data across time for each unit of analysis. You may consider merging data into a wide format for each participant level (e.g., student-level dataset, teacher-level dataset, school-level dataset), where all forms associated with each case are found in one row. See Section 3.3.2 for more information on how to combine files in this way.
- Separate files
  - On the flip side, sharing separate files may reduce burden for researchers interested in individual datasets. For instance, if a user is only interested in one instrument collected from one participant group, having the datasets separated allows researchers to download just the file for that one instrument, rather than downloading a larger dataset and dropping variables not relevant to their research.
  - If sharing longitudinal datasets separately, decide now if you want to go ahead and add time components to your variable names (e.g., `w1_`, `w2_`), or add a time component variable (e.g., `wave`) to your datasets, removing the need for future users to have to do this step when combining data (see Section 9.4.1 for more information on adding time to your dataset).

- When sharing datasets separately, also consider developing a folder and file naming structure that allows researchers to easily know what files they are working with. Also make sure that all files have identifiers (i.e., keys) necessary to link datasets. Last, you'll want to include a README for future users that helps clarify how datasets can be combined (refer to Section 8.3.1 for more information).

16.2.3.3 File Formats

Most funders require data to be shared in an electronic format, and for quantitative data in particular, that usually means a rectangular format. In keeping with FAIR principles (Section 2.4.1), it is recommended to provide data in at least one non-proprietary format (see Table 15.1 for example formats). As discussed in Chapter 13, this not only allows a broader audience to access your data, but also protects against technological obsolescence.

However, as covered in Chapter 14, it can be beneficial to also share data in formats that have embedded metadata (e.g., SPSS, Stata). Providing your data in both a non-proprietary format and a proprietary format that is widely used in the field, can give your users options while also protecting your data from obsolescence (Institute of Education Sciences 2023a; Neild, Robinson, and Agufa 2022).

Most importantly, though, if sharing in a repository, check to see if there are any data format requirements. ICPSR, for example, encourages submission of files with embedded metadata, such as SPSS, Stata, or SAS files (ICPSR 2023b). They then use these files and curate them into ASCII data with setup files to accompany statistical programs. As another example, the National Archive of Criminal Justice Data used by the National Institute of Justice (NIJ) prefers SPSS formats for quantitative data, but will also accept Stata or SAS files (National Archive of Criminal Justice Data 2023).

16.2.3.4 Assess Disclosure Risk

Before publicly sharing study data, it is imperative that you conduct a disclosure risk assessment. In conducting this assessment, review variables that could potentially identify a participant, either directly or indirectly, and also review sensitive variables that have the potential to cause harm to participants if their identity is disclosed.

1. Direct identifiers
    - As discussed in Chapter 4, these are identifiers that are unique to an individual and can be used to directly identify a participant (e.g., name, email, IP address, student ID). It can be helpful to mark these identifiers in your data dictionary early on to keep

track of what should be removed. If you are unsure exactly what direct identifiers to check for, the 18 protected health identifiers listed in the HIPAA safe harbor de-identification method[3] are a good starting point. FERPA[4] also provides a list of personally identifiable information (PII) to review.

2. Indirect identifiers
   - While our dataset may be technically de-identified after removing our direct identifiers, it is still important to consider the possibility of deductive disclosure (Institute of Education Sciences 2023a). Research has shown that it is possible to re-identify someone through a combination of indirect identifiers (Sweeney 2002) (see Table 4.1 for examples). Further care must be taken to consider if there are any other ways participants can potentially be re-identified in our data. This includes considering the following:
     - Open-ended questions: These variables may contain information that can directly or indirectly identify individuals.
     - Outliers: If someone has extreme values for a variable, it may be easier to identify that individual.
     - Small cell sizes: If there is only one person who took a survey on a particular date, or only one person who fits in a specific demographic category, it is easier to re-identify that individual. The NCES Standard 4-2-10, suggests that all categories have at least three cases to minimize risk (Seastrom 2002), while others may recommend more stringent requirements such as a minimum of five cases (Schatschneider, Edwards, and Shero 2021).
       - Combinations of variables, or crosstabs, can also create small cell-sizes (e.g., a student may be identifiable by school size + special education status + gender + grade level). Generally, the more indirect identifiers you have in your dataset, the more possible combinations exist, increasing the risk of re-identification (Morehouse, Kurdi, and Nosek 2023).
   - When reviewing this information, consider not only information that the general public may be able to decipher, but also what information may be known to people who know a participant (e.g., administrator, teacher, parent). You also want to consider the amount of potentially publicly available information about a participant or site (e.g., administrative datasets, social media data) and the likelihood that public information could be used to re-identify someone (i.e., by linking public data with your study data) (Filip 2023; Meyer 2018; Neild, Robinson, and Agufa 2022).

3. Sensitive information
    - In assessing disclosure risk, you also want to review any variables that could cause potential harm to an individual if they were re-identified. Examples of these variables include health information, special education status, disciplinary status, or information on risky behaviors (Morehouse, Kurdi, and Nosek 2023; Neild, Robinson, and Agufa 2022).

16.2.3.4.1 Mitigating Disclosure Risk

De-identification is a process of balancing risks and benefits, and those risks and benefits should be equally distributed across participant groups in your data (McKay Bowen and Snoke 2023). In assessing disclosure risk, you need to weigh the level of potential harm that may be caused if a participant's identity is uncovered in your data (e.g., legal repercussions, embarrassment) against the potential benefits incurred by sharing data (e.g., advancing science). Even after editing data, the amount of risk is never zero (Neild, Robinson, and Agufa 2022). It is always possible that a user can find a way to re-identify someone through combining variables within your project datasets, or by linking your project data to other publicly available information. During the de-identification process, you want to decrease risks of re-identification without seriously reducing the utility of the dataset (e.g., consider reproducibility of findings).

As discussed in Chapter 14, first, direct identifiers should be completely removed from each dataset and replaced with your study IDs. By the time you get to the point of preparing for data sharing, this step should hopefully already be completed. Yet, there may be more to consider. As an example, to further reduce participant re-identification risk, even if schools and districts were not necessarily promised confidentiality, it is beneficial to remove names and replace those locations with unique study IDs. Furthermore, if you have any concerns about the confidentiality of your unique study IDs (e.g., schools have seen the list of identifiers and can associate participants to individual IDs), you may want to consider assigning a new set of IDs in your data before publicly sharing. If you do this, make sure to track both sets of identifiers in your participant tracking database.

Next, consider what indirect identifiers and sensitive variables exist in your data. When dealing with these types of variables there are a variety of commonly used methods for reducing disclosure risk. Examples of methods that can be easily applied (i.e., do not require any special technical expertise) and do not change the underlying values in the data are listed below (Filip 2023; Garfinkel 2015; Logan, Hart, and Schatschneider 2021; Morehouse, Kurdi, and Nosek 2023; Neild, Robinson, and Agufa 2022; Schatschneider, Edwards, and Shero 2021).

- Redaction: Eliminate the entire variable from the data.
    - This method should be used for all direct identifiers and may also be used for indirect identifiers that pose disclosure risks.

  - Important indirect identifiers to consider are open-ended text variables with verbatim responses. Although identifiers, such as name, should have been removed from these variables in the data cleaning phase (see Section 14.3.1), participant responses are still unique and potentially identifiable, increasing the risk of disclosure. It's important to review responses to determine the level of risk. If your team determines that an open-text variable should be redacted, researchers can still maintain information by coding common responses and creating a new categorical variable. This variable, and the coding scheme, should be documented for future users.

- Suppression: Remove data in a particular cell or row.
  - You can either leave the cell as missing or fill with a code to indicate the value has been suppressed.
- Generalization: Reduce precision in the data. This includes techniques such as:
  - Reporting a range as opposed to distinct values (e.g., range of years teaching as opposed to number of years).
  - Using rounded values rather than exact numbers.
  - Collapsing categories (e.g., creating an "other" category for all special education categories with Ns < 5).
  - Creating summary variables (e.g., use `date of data collection` and `date of birth` to create an age variable, allowing you to remove both `date of birth` and `date of data collection`).
  - Report in larger units (e.g., reporting `age` in years as opposed to `age` in months, or reporting state or region-level `geography` as opposed to county or district).
    - For geography in particular, the HIPAA safe harbor method recommends removing all geography smaller than state (U.S. Department of Health and Human Services 2012).
- Truncation: Also called top or bottom coding, this involves restricting the upper and lower ranges to mask outliers (e.g., top code any `income` above $150k to "$150k or higher").
- Share unlinked files: If you find that the potential for re-identification is caused when linking across files (e.g., linking a student file to a teacher file), and the indirect information contained in those files is necessary for analysis purposes, you may consider sharing a set of files that do not contain linking variables. However, this is not ideal as it limits future use cases for your data.

In most cases, these easily applied methods should be satisfactory. However, in situations where the risk of disclosure is more than minimal and the level of potential harm is also more than minimal, you may consider using more advanced techniques. Since these techniques require more technical expertise and have the potential to drastically alter your data if not applied correctly, it is important to consult with someone who has expertise in these methods before attempting them (e.g., methodologists, repository curators, research data librarians). A few of these advanced techniques include the following.

- Swapping: Matching cases on one or more key variables, then swapping values for the indirect identifiers of interest.
- Perturbing: Adding random statistical noise to the data (e.g., multiply all values of a variable by a random number).
- Microaggregation: Replace an individual's value in a cell with the average value of their small group.

The process of de-identifying data before data sharing should not be done alone. Schedule one or more meetings with your data management working group to review data and develop a plan. Bring in outside expertise as needed, especially when implementing more advanced data de-identification techniques. Last, it is imperative that you document your de-identification methods in the appropriate locations (e.g., research protocol, data cleaning plan, data dictionary) so that you and future users can track all processing done to your data.

Resources

Source	**Resource**
Alena Filip	Table 2 provides pros and cons of various de-identification methods[5]
J-PAL	Table 3 provides a list of direct and indirect identifiers and recommended removal methods[6]

16.2.3.4.2 Sharing Controlled Access Data

If you find that de-identifying your data alters it in a way that distorts data quality or structure, or if you believe that the risks of sharing your data are still more than minimal, you should consider sharing in a controlled manner (ICPSR 2020; Meyer 2018; Morehouse, Kurdi, and Nosek 2023; Neild, Robinson, and Agufa 2022; Schatschneider, Edwards, and Shero 2021). As discussed in Section 16.2.1, it is still possible to share these files through restricted access. In a repository these files can be shared through means such as secure downloads, virtual data enclaves, or even onsite data enclaves. Access to these restricted data is permitted only through an application process where requestors complete a detailed data use agreement.

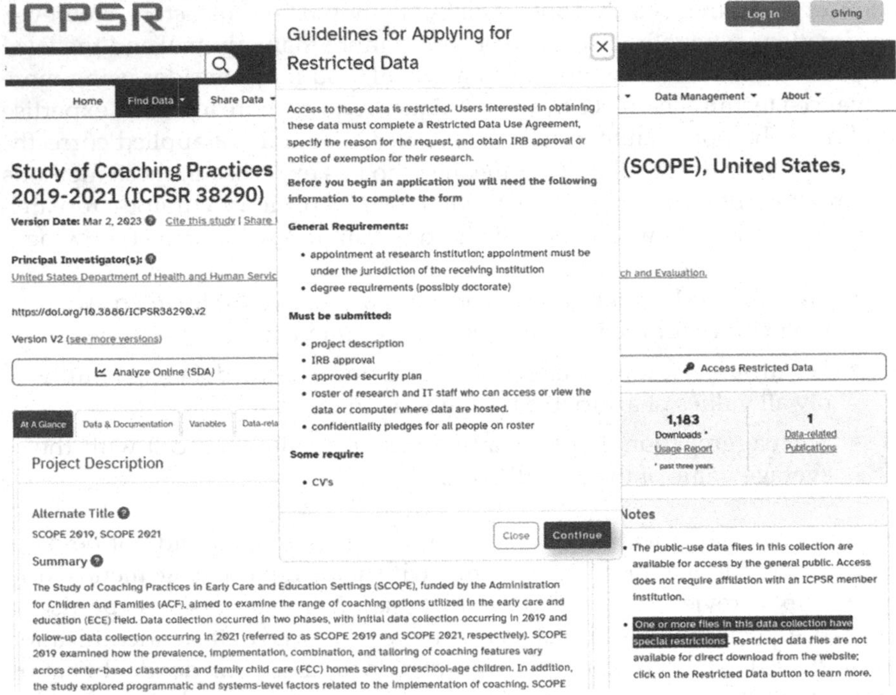

FIGURE 16.4
An example of restricted access data on ICPSR.

See Figure 16.4 for an example of a restricted access dataset available from the United States Department of Health and Human Services. Administration for Children and Families, Office of Planning, Research and Evaluation (2023) in the ICPSR data repository.

If you are using a repository that does not accept restricted data, you can still share metadata in a repository, along with information for users to contact you about requesting access to restricted-use data that you share through your own personal system. No matter where you share data, in order to maximize public benefit, you should still consider openly sharing some data. That may simply be summary statistics (e.g., means, standard deviations) provided in tables. Or it could involve sharing two sets of files. For example, a public access version of a file with sensitive variables removed/altered and a restricted-use version of the same file with sensitive variables retained. However, be sure to consider all possible disclosure risks before sharing to ensure that someone with access to both the restricted and public files are not able to identify individuals. Also make sure that no inconsistencies between files are created during this process (ICPSR 2020; Logan, Hart, and Schatschneider 2021; Neild, Robinson, and Agufa 2022; Schatschneider, Edwards, and Shero 2021).

16.2.4 What Documentation to Share

In the 2022 OSTP memo (Nelson 2022), it is stated that all federal agencies will be expected to develop data sharing policies that elicit free and open access to scientific data that is of "sufficient quality to validate and replicate research findings". As we've learned throughout this book, sharing data alone is not sufficient to enable data reuse (Hardwicke et al. 2018). It must be accompanied by thorough documentation to allow a user to understand data provenance. In considering data documentation, you should not only consider the amount of documentation, but also the quality. The higher quality of documentation you provide, the higher likelihood that your data will be reused (Goodman et al. 2014).

In deciding exactly what documents to share, check with your funder to see if specific documents are required. If you are depositing your data in a repository, check what is required there as well. Each repository may mandate or suggest types of documentation to provide. ICPSR, for example, provides suggested documentation to include, such as codebooks, instruments, README files, project summary documents, and publications (ICPSR 2023b).

You will want to share documents from all levels of your project (project, data, and variable). Each level of documentation will provide unique contributions to help users understand the background of your project, how files are related, and how to interpret and use variables. Ideas of what to share from each level are provided (see Chapter 8 for more details about each of these documents).

1. Project-level documentation
   - Research protocol
     - This document, which may be called other names (e.g., project summary document[7]), provides the what, who, when, where, and how of your study. It is the summation of everything that occurred in a study and provides users with all of the background knowledge necessary to correctly interpret and use your data.
     - Along with your research protocol, you can also share a variety of other helpful supplemental documents including, but not limited to:
       - Participant flow diagrams
       - Timelines
       - Copies of data collection instruments
       - Copies of consent forms
   - Project-level README
     - A README can serve many purposes, and in this case, a README at the top of your data sharing directory can be

beneficial in outlining high-level information about the project (i.e., title, overview, contributors), as well providing a file tree that describes how files are organized in the directory.

An example of a project-level README file.

```
Project title and subtitle

Brief project overview
Describe this project or resource.

Contributors
List all contributors.

Repository overview
Provide an overview of the directory structure and files,
for example:

study_name/
|-- 00_README.txt
|-- documentation
    |-- study_name_project_summary_document.pdf
    |-- study_name_stu_svy_data_cleaning_plan.txt
    |-- study_name_stu_svy_data_dictionary.xlsx
|-- data
    |-- 00_README.txt
    |-- study_name_w1_stu_svy_clean.csv
    |-- study_name_w2_stu_svy_clean.csv
|-- analysis
    |-- 00_README.txt
    |-- data
    |-- manuscripts
    |-- output
    |-- syntax

Applicable instructions
Any instructions/tools necessary to use files or replicate
results.

Additional resources
Point interested users to any related literature and/or
documentation.
```

2. Dataset-level documentation
    - Dataset-level README
        - In this case, a README organized in a rectangular or other format can be very helpful to describe a set of files in a specific

folder (e.g., a student data folder). This README can list out every file in the folder and provide information that allows users to understand what each file contains as well as how datasets may be linked. This is especially beneficial in cases where you are sharing individual files with the expectation that future users will combine them on their own time. Adding this README clarifies what identifiers can be used to link files (see Figure 8.13 for an example).

- Syntax files
    - In most cases, it is unlikely you are going to want to publicly share your data cleaning code. For one, you are typically not sharing the raw data associated with your cleaning code, so your cleaning code is not useful for reproducibility purposes. Second, depending on the amount of data you collected and the amount of cleaning that was required for your data, cleaning code can be overwhelming for a future user to sift through. Sometimes it can be 20 or more scripts if you have multiple instruments, collected across multiple participant groups, and multiple waves. Most likely this is not helpful for anyone, outside of just learning more about your coding practices (which can actually be very helpful for teaching and learning purposes). In most cases, though, it is not necessary to share your data cleaning code unless you really want to. However, if you plan to share analytic files in your repository, it is imperative that you share any code associated with the creation of those analytic datasets as well as any code necessary to reproduce findings (Renbarger et al. 2022).
- Data cleaning plans
    - While it is not always necessary to share your data cleaning code, especially when you are not sharing your raw data, it can be helpful to share your data cleaning plans. Providing these plans allows some transparency into the decision-making process that occurred when transforming your raw data into your clean shareable data, which may help future users with interpretation.

3. Variable-level documentation
    - Data dictionaries and/or codebooks
        - These documents help users correctly interpret variables in your dataset. At least one of these documents must be shared, if not both. Recall that data dictionaries provide an overview of the variables that exist in each dataset, displayed in a tabular format (Section 8.4.1). Codebooks provide somewhat similar

information, in addition to summary statistics about each variable, and are usually provided in text format (Section 8.4.2).

4. Metadata
   - As discussed in Section 8.5, if sharing data in a repository, metadata will most likely be collected by the repository when you deposit the materials. While metadata can be added for all levels, at minimum your repository will most likely collect project-level metadata (e.g., title, creator, date, key words). This machine-readable metadata will aid in the discoverability and reusability of your materials.

Before sharing any documentation, make sure you have assessed all forms for disclosure risk. Review code, data dictionaries, screenshots, and so forth, for any instances where names or other identifying information is used and redact all PII before sharing (Neild, Robinson, and Agufa 2022). It is also important to consider any copyright issues you may encounter. If a published measure does not give you permission to share item-specific information, you may need to remove any exact verbiage from your documentation (e.g., data dictionaries, codebooks) and replace with generic language such as "item 1 of assessment name" (Logan 2021).

16.2.4.1 File Formats

Again, if you are sharing data in a repository, check to see if they have required or suggested documentation formats. For example, ICPSR specifies a variety of formats for documentation submission including Microsoft Word, ASCII, or DDI XML format and they in turn convert your documentation to XML and PDF formats (ICPSR 2023b). You'll also want to check if there are any metadata standards your repository complies with, such as DDI standards, and if that requires any additional considerations on your part or if the archive takes care of standardizing information for you.

If not depositing in a repository, or if your repository has no strict requirements, sharing documentation in non-proprietary formats is recommended. These are files anyone can open no matter what software they own. PDFs and text files are generally good formats. If a rectangular format is used, for documents such as a data dictionary, CSV files can be created and still opened in any spreadsheet program. Some researchers have also started to share documentation in searchable formats, including HTML, to make it easier to sift through large amounts of information.

16.2.5 When to Share

The most important thing to consider here is what is required by your funder, or other supporters. Federal funders will likely have required time frames. As an example, NIH expects that researchers will share their data no later than at the time of publication or by the end of their grant, whichever comes first (National Institutes of Health 2021). While other funders, such as the National Institute of Mental Health (NIMH), have expectations that grant awardees will share data throughout the grant on a regular schedule (National Institute of Mental Health 2023).

Your repository may also have preferences for when data should be deposited. For example, while data can be deposited at any time with ICPSR, they recommend beginning uploading data as soon as possible after data collection to allow time for data curation (ICPSR 2023c). If you choose to make ongoing deposits of data, you will want to consider if you would like to embargo your data until you are required to share. This means you delay the public release of your data until a later time point of your choosing. This allows you to continually deposit your data, for example to reduce the workload at the end of your study, while also allowing you sole access to your data while your project is active. If you choose to deposit data on an ongoing basis, make sure that your repository allows you to edit, update, and version your data if errors are found or changes are needed at any point during the project.

16.3 Repository File Structure

Some repositories have a predefined file structure and will require little to no effort on your part to set this up. However, many repositories are more hands off and will allow you to set up structures in any way that works well for you. Similar to how you followed a style guide (Chapter 9) for setting up your internal team and project electronic file structures, you will want to follow style guide rules when setting up your data sharing file structure. Be organized about how you store information and be descriptive in how you name folders and files. You want to remove as many cognitive barriers as possible to increase the likelihood that your materials are reused. If someone opens your project and they feel confused at any point about what they are looking at, it is likely they will leave and move on to a new project. Users want to feel confident that they understand what is in your project and how they should use the materials. Also make sure to use any additional services provided by your repository that may also aid discoverability and interpretation (e.g., wikis, linking to other relevant materials).

Here is an example of one way you might set up a repository for a small project that collected longitudinal data for both students and teachers. The "analysis" folder would only be added if, in addition to sharing general clean study files, you also plan to include any files created for specific analyses. If adding an "analysis" folder, it can be helpful to include another project-level README at the top of the folder, describing any helpful information about the specific analysis, as well as an overview of how that folder is organized.

```
study_name/
├── 00_README.txt (project-level)
├── 01_documentation
|   ├── data_cleaning_plans
|   |   ├── study_name_stu_svy_data_cleaning_plan.txt
|   |   └── study_name_tch_svy_data_cleaning_plan.txt
|   ├── study_name_project_summary_document.pdf
|   ├── study_name_stu_svy_data-dictionary.csv
|   ├── study_name_tch_svy_data-dictionary.csv
├── 02_data
|   ├── stu
|   ├── 00_README.txt (dataset-level)
|   |   ├── w1
|   |   |   ├── study_name_w1_stu_svy_clean.csv (CSV
|   |   |       format)
|   |   |   └── study_name_w1_stu_svy_clean.sav (SPSS
|   |   |       format)
|   |   ├── w2
|   |   |   ├── study_name_w2_stu_svy_clean.csv
|   |   |   └── study_name_w2_stu_svy_clean.sav
|   ├── tch
|   ├── 00_README.txt (dataset-level)
|   |   ├── w1
|   |   |   ├── study_name_w1_tch_svy_clean.csv
|   |   |   └── study_name_w1_tch_svy_clean.sav
|   |   ├── w2
|   |   |   ├── study_name_w2_tch_svy_clean.csv
|   |   |   └── study_name_w2_tch_svy_clean.sav
├── 03_analysis (if relevant to this repository)
|   ├── 00_README.txt
|   |   └── …
|   ├── data (this would contain analytic data files)
|   |   └── …
|   ├── manuscripts
|   |   └── …
|   ├── outputs
|   |   └── …
|   ├── syntax
|   |   └── …
|
|____
```

Resources	
Source	**Resource**
Crystal Lewis	Example education project data shared in the OSF data repository[8]

16.4 Roles and Responsibilities

As with every phase in the research life cycle, it is so important to assign roles and responsibilities throughout this process. Make a checklist of everything that needs to happen in this data sharing process and assign someone to every task. Many of the required tasks (e.g., data cleaning, documentation) should not require additional planning, as those roles and responsibilities are designated in other phases. However, there are several new tasks that will be specific to this phase (e.g., creating data use agreements, assessing disclosure risk), and some tasks will vary depending on if you are sharing through a repository/institutional archive (e.g., communicating with repository staff, troubleshooting issues) or sharing on your own (e.g., developing a data request and sharing system). Even after data has been shared, depending on how and where your data is shared, you may also need to assign responsibilities for ongoing maintenance (e.g., responding to access requests).

As you begin assigning roles, you may discover that additional expertise is needed, especially for things like the data de-identification process or developing data sharing agreements. If yes, also begin looking into who may fill these gaps (e.g., research data librarians, methodologists, data curation specialists at your repository). Once your data sharing plan is formalized, it should be documented in all necessary locations (e.g., DMP, research protocol, SOPs, informed consent forms).

A few organizations have put together data sharing checklists. These checklists may help you begin assigning team members to specified responsibilities.

Resources	
Source	**Resource**
Data Curation Network	A standardized set of steps and checklists for reviewing datasets before publicly sharing[9]
Institute of Education Sciences	Data sharing checklist on page 30[10]

16.5 Revisions

When publicly sharing study data and materials, researchers are often concerned that errors may be found (Beaudry et al. 2022; Houtkoop et al. 2018; Levenstein and Lyle 2018). If errors are found in your data after publicly sharing, while not ideal, it is important to recognize that you are not the first person this has happened to. Other researchers have come across errors in their data after sharing and have successfully made plans for addressing the problem (Aboumatar et al. 2021; Grave 2021; Laskowski 2020; Strand 2020). Depending on how you've shared and used your data, there are different ways to address these errors.

First, if you have used your data in a publication, contact your journal to make them aware of the errors you found. The journal may provide options, including the opportunity to revise or retract your article. Next, you will want to update your shared data.

If your data is shared in a repository or institutional archive:

1. Make the appropriate edits to your data and upload the new version to your repository.
   - Many repositories will then assign your project, or data, a new version number, along with a new DOI, to denote the changes in the project.
   - If the repository does not provide a place to note the reason for revisions, add this information in your own changelog or README in the project folder (Kopper, Sautmann, and Turitto 2023b; Towse, Ellis, and Towse 2021).
2. If the repository requires/allows users to make an account before accessing the data, they may have a system to email current users to let them know that a new version of the data has been created.

If your data is not in a repository or institutional archive:

1. Make the appropriate edits to your data and save as a new version. Make a note about the change in your internal changelog.
2. Then, consider personally reaching out to anyone who has submitted a data request to notify them of the errors in the data.

While the potential for others to find errors is often viewed as a consequence to data sharing, we can also view this as an opportunity to do better science. Giving others the chance to catch errors in our data that we may not have otherwise caught (Bishop 2014; Klein et al. 2018; Schoen and Solmaz-Ratzlaff

2023) allows us to make corrections, ensuring that our findings are not derived based on inaccurate data. Data sharing also provides incentive for us to implement more rigorous data management practices that hopefully improve data integrity and create less concern for future errors (Klein et al. 2018; Strand 2021).

Notes

1 https://ies.ed.gov/ncee/pubs/2022004/pdf/2022004.pdf
2 https://zenodo.org/records/7946938
3 https://www.hhs.gov/hipaa/for-professionals/privacy/special-topics/de-identification/index.html#safeharborguidance
4 https://www.ecfr.gov/current/title-34/subtitle-A/part-99
5 https://www.sjsu.edu/research/docs/irb-data-management-handbook.pdf
6 https://www.povertyactionlab.org/resource/data-de-identification
7 https://osf.io/q6g8d
8 https://osf.io/59gte/
9 https://datacurationnetwork.org/outputs/workflows/
10 https://ies.ed.gov/ncee/pubs/2022004/pdf/2022004.pdf

17

Additional Considerations

Up until this point, we have mostly focused on a workflow for one team with one project. However, it is common for projects to include multi-site collaborations (e.g., a grant shared across multiple research institutions), and for teams to have multiple projects. Both of these add complexity to data management which I will briefly address.

17.1 Multi-Site Collaborations

Multi-site, or multi-team, collaborations require additional planning around roles and responsibilities, workflows, and standards. Jumping into multi-site collaborations without spending time cross-team planning often leads to unfortunate data security, quality, and usability concerns. Before a project begins, consider documenting expectations in a collaboration agreement. When developing this type of agreement, review everything in a typical data management checklist but come to an agreement on decisions. The following types of multi-site issues should also be addressed (Briney 2015; Schmitt and Burchinal 2011).

- How will teams maintain consistency in procedures across sites (e.g., shared SOPs, shared style guides, oversight of practices)?
  - If each site is handling its own data tracking, capture, entry, and cleaning procedures, it is imperative that these processes are standardized to allow for datasets to be integrated.
- How will teams handle data ownership?
- What are the roles and responsibilities across sites?
- What tools will be used to allow for multi-site data tracking, collection, entry, storage, and sharing?

Documents, such as a RACI matrix (Miranda and Watts 2022), can help lay out expectations for team collaborations (Figure 17.1). In these charts, each site is assigned to a task as either responsible, accountable, consulted, or informed.

DOI: 10.1201/9781032622835-17

Project Tasks	Site A	Site B	Site C
Develop protocol	C	R	C
Develop data dictionaries	R	R/A	R
Develop tracking tool	R	A/C	I
Develop data collection tools	R	A/C	I
Collect data	R	R	R
Capture/Clean data	I	R	I

FIGURE 17.1
Example of a simplified RACI chart for a multi-site project collaboration.

- Responsible: The site is responsible for completing this task.
- Accountable: The site provides oversight, ensuring the task is completed with fidelity.
- Consulted: The site is always consulted before a decision is made on this task.
- Informed: The site is provided high-level information about decisions that are made.

Assigning levels of responsibility to each site allows collaborators to clearly see what is expected from them in a project. Within site tasks can then be further assigned to specific roles.

17.2 Multi-Project Teams

Similar to multi-site projects, organizing multiple projects within a team requires additional coordination. As your number of active projects grows, the sophistication of your operations should grow as well. Consider doing the following (Van den Eynden et al. 2011):

1. Centralize resources.
    - Create templates, SOPs, code snippets (reusable blocks of code), and style guides that can be used across projects. Utilize your team wiki to post shared resources in a central location where team members can easily access them (see Section 8.1.2). Centralizing resources reduces duplication of efforts, and also improves standardization, allowing you to more easily integrate data across projects.

2. Encourage team science.
    - When running multiple grants, the "lone cowboy" model of having just one person manage everything becomes even less feasible (Reynolds et al. 2014). Embrace the idea that it takes a team of people, skilled in many different areas (e.g., project management expertise, data management expertise, content expertise, administration expertise), to do quality research (Bennett & Gadlin 2012). With more than one grant, it is potentially more feasible to hire people to fill specialized roles, and to fund them across multiple grants.
3. Create a hierarchy of roles.
    - As the size of a team increases, it becomes important to assign someone to foster collaboration and oversee fidelity to data management standards across projects (Briney 2015). Create a hierarchy of roles, including data management implementers (e.g., data specialists, data managers) and supervisors (e.g., senior data managers, data leads), with the supervisor role helping to prevent internal drift through expectation setting, oversight, and mentorship.
4. Create support systems.
    - If your team is large enough, and you have multiple people working on data management across different projects, it may be helpful to create a data core. This internal group of data people can meet regularly to share knowledge and resources, develop and modify shared documentation, and develop internal data trainings for staff, increasing capacity for your center.

17.3 Summary

Collecting data is a bit like cooking a good meal. If you clean as you go, when you are full and sleepy you will have much less to do.

– Felicia LeClere (2010)

Slow science is often used to describe an antithesis to the increasingly fast pace of academic research, instead suggesting that science should be a slower, more methodical process (Frith 2020). Likewise, if we hurry a research project along without spending time putting quality data management processes into place, we increase the possibility that we end up putting research into the world that we cannot trust. Instead, we should take time to plan data management before a project begins, and implement quality practices throughout the life cycle.

While it may be difficult to support this slow process early on, remember that data management gets easier the more you do it. Once you have templates, protocols, and style guides in place, those documents and processes can be reused, easing burdens in future projects (Levenstein and Lyle 2018).

There is no one-size-fits-all approach to data management though (Bergmann 2023; Reynolds et al. 2014). Projects are nuanced, and there is no way to anticipate every way in which each specific project's data needs to be managed. Instead, use the "buffet approach" and implement what works best for your project and your team (Bergmann 2023). What matters most is that those practices are implemented consistently within your project, and that ultimately they produce quality, well-documented, data products that are accepted in the field. Similarly, while it is possible that all of the practices mentioned in this book work for your project, it is unlikely that your team has the bandwidth to do it all. Instead, implement "good enough" practices that allow you to achieve the quality outcomes you desire (Borghi and Van Gulick 2022; Wilson et al. 2017). You don't have to create all the documentation or use the most sophisticated data cleaning methods. You simply need to use methods that are good enough to reach your goals. Also make sure to periodically review your data management practices to ensure you are keeping up with changing requirements, technologies, standards, or team/project needs. Consider holding a data management retreat or workday once a year in order to have time set aside to review procedures as a team (Spinks 2024).

Last, as the awareness of the necessity of good data management grows in our field, we can only hope that systemic changes will continue to happen, making these efforts easier for education researchers. While institutions, such as academic libraries, provide data management learning opportunities including workshops, online modules, and on-demand courses, these resources are still not reaching everyone. Integrating data management content into required college coursework would improve data management practices for all the researchers who are out there "winging it" because they learned data management through informal methods. Funding institutions may also begin to find ways to necessitate data management training for applicants. With requirements for data management and sharing expanding, Wilson et al. (2017, 19) suggest that "it is unfair as well as counterproductive to insist that researchers do things without teaching them how". Furthermore, developing shared standards for the field of education would do wonders in easing the burden on researchers who are having to make on the fly data management decisions, instead allowing them to follow a set of instructions for tasks such as formatting and documenting their data. Developing these standards would also benefit anyone interested in scientific inquiry, improving consistency in the quality and usability of publicly shared data products.

Glossary

Term	Other Terms	Definition
Anonymous data	NA	Identifying information was never collected. This data cannot be linked across time or measures.
Aggregated data	NA	Individual data that is summarized at a group level.
Append	Vertical join, join columns, union	Stacking datasets on top of each other (matching variables).
Archive	NA	The transfer of data to a facility, such as a repository, that preserves and stores data long-term.
Attrition	NA	The loss of study units from the sample, often seen in longitudinal studies.
Clean data	Processed data	Raw data that has been manipulated or modified for the purposes of correcting and clarifying information.
Cohort	NA	A group of participants recruited into the study at the same time.
Coded data	Pseudonymized data, indirectly identifiable, confidential data	Personally identifiable information (PII) has been removed and names are replaced with a code. The only way to link the data back to an individual is through that code. The identifying code file (linking key) is stored separate from the research data.
Confidential data	NA	This data is protected from unauthorized disclosure. This data either contains personally identifiable information or can still be linked back to an individual through other means (e.g., identifiable data or coded data).
Confidentiality	NA	Confidentiality concerns data, ensuring participants agree to how their private and identifiable information will be managed and disseminated.
Control	Business as usual (BAU)	The individual or group does not receive the intervention.
Cross-sectional	NA	Data is collected on participants for a single time point.
Data	Research data	The recorded factual material commonly accepted in the scientific community as necessary to validate research findings (OMB Circular A-110).

(*Continued*)

Term	Other Terms	Definition
Data repository	Data archive	A storage location for researchers to deposit data and supporting materials associated with their research.
Data structure	NA	A way of organizing data to allow for more efficient processing and storage. In particular, repeated measures data can be structured in either long or wide format.
Data type	Measurement unit, variable format, variable class, variable type	A classification that specifies what types of values are contained in a variable and what kinds of operations can be performed on that variable. Examples of types include numeric, character, logical, or datetime.
Database	Relational database	An organized collection of related data stored in tables that can be linked together by a common identifier.
Database design	Database schema, data modeling	A collection of decisions regarding how tables, or datasets, will be organized and related to one another.
Dataset	Data set, data frame, spreadsheet, rectangular data, tabular data, table	A structured collection of data usually stored in tabular form. A research study may produce one final dataset per entity/unit (e.g., teacher dataset, student dataset).
De-identified data	Anonymized data	Identifying information has been removed or distorted and the data can no longer be re-associated with the underlying individual (the linking key no longer exists).
Derived data	Calculated values	Data created through transformations of existing data (e.g., mean scores).
Direct identifiers	NA	These variables are unique to an individual and can be used to directly identify a participant (e.g., name, email address).
Directory	File structure, file tree, folder structure	A cataloging structure for files and folders on your computer.
Disclosure risk	NA	The risk of re-identifying a participant and the harm that may come from that disclosure.
Extant data	Secondary data, administrative data, third-party data, external data	Existing data generated/collected by external organizations at an earlier point in time (e.g., school records).

(*Continued*)

Term	Other Terms	Definition
FERPA	NA	The Family Educational Rights and Privacy Act is a federal law governing the disclosure of personally identifiable information in education records (e.g., name, address, DOB). The law applies to all public elementary and secondary schools, as well as post-secondary institutions.
File format	File type, file extension	A way that information is encoded for storage on a computer. There are both proprietary (e.g., SPSS, XLSX) and non-proprietary formats (e.g., CSV, TXT).
Foreign key	NA	One or more variables associated with unique values in another table
Human subject	NA	The Common Rule (45 CFR 46) definition of a human subject is a living individual about whom an investigator conducting research obtains: (1) Data through intervention or interaction with the individual, or (2) identifiable private information.
HIPAA	NA	The Health Insurance Portability and Accountability Act is a federal law covering the protection of sensitive health information.
Identifiable data	NA	Data that includes personally identifiable information.
Indirect identifiers	Quasi-identifiers	These variables do not alone identify a particular individual (e.g., ethnicity, gender), but, if combined with other information, they could be used to identify a participant
Instrument	NA	A mechanism designed to collect original data (e.g., observation form, questionnaire, assessment)
Longitudinal data	Repeated measures	The same information is collected from the same subjects at multiple time points.
Measure	Scale	In this book, I use the term "measure" broadly to refer to a collection of items used to measure an outcome (e.g., an existing scale, an existing academic assessment).
Merge	Horizontal join, join rows, link	Combining datasets together in a side-by-side manner (matching on one or more unique identifiers).
Metadata	NA	Data providing details about other data.

(Continued)

Term	Other Terms	Definition
Missing data	NA	Occurs when there is no data stored in a variable for a particular observation/ respondent.
Normalize	NA	In this book, the term "normalize" is used to refer to returning a value to its normal, or expected state.
Original data	Primary data	Firsthand data that are generated/collected by the research team as part of the research study.
Participant database	Study roster, master list, master key, linking key, tracking database	This database, or spreadsheet, includes any identifiable information on your participants as well as their assigned study ID. It is your only means of linking your confidential research study data to a participant's true identity. It is also used to track data collected across time and measures as well as participant attrition.
Path	File path	A string of characters used to locate files in your directory system.
Persistent identifier	PID	A unique and enduring digital reference to an object, contributor, or organization. A DOI (digital object identifier) is a type of PID specific to digital objects.
Personally identifiable information	PII, personal data	This includes direct identifiers (e.g., name and email), as well as indirect identifiers that, if combined with other variables or if in small enough numbers, could identify a participant (e.g., full birthdate and place of birth).
Primary key	NA	One or more variables that uniquely define rows in your data
Privacy	NA	Privacy concerns people, ensuring they are given control to the access of themselves and their information.
Private data	NA	Highly restricted and typically not publicly shared, or is shared with limited access (i.e., passwords, illegal behaviors, medical records, financial information).
Protected health information	PHI	The HIPAA Privacy Rule provides protections for 18 identifiers held by covered entities providing health care services.
Qualitative data	NA	Non-numeric data typically made up of text, images, video, or other artifacts.
Quantitative data	NA	Numerical data that can be analyzed with statistical methods.

(Continued)

Term	Other Terms	Definition
Randomized controlled trial	RCT	A study design that randomly assigns participants, or groups of participants, to a control or treatment condition.
Raw data	Primary, untouched	Unprocessed data collected directly from a source.
Replicable	NA	Being able to produce the same results if the same procedures are used with different materials.
Reproducible	NA	Being able to produce the same results using the same materials and procedures.
Research	NA	The Common Rule (45 CFR 46) definition of research is a systematic investigation, including research development, testing, and evaluation, designed to develop or contribute to generalizable knowledge.
Restricted-use data	Non-public data, controlled data, managed access data	A dataset that cannot be publicly released due to containing sensitive information or a combination of variables that could enable identification. These data require controlled access conditions and may be shared through data use agreements or other application processes.
Safe harbor method	NA	Under the HIPAA Privacy Rule, there are two methods of de-identification. The safe harbor method allows covered entities to treat data as de-identified if all 18 PHI variables are removed.
Sensitive data	Protected data	An umbrella term that encompasses proprietary, ethical, contractual, or private information that should be protected from unwarranted disclosure. There are varying levels of data sensitivity.
Standardize	NA	Developing and implementing a set of consistent procedures
Study	NA	A single funded research project resulting in one or more datasets to be used to answer a research question.
Subject	Case, participant, site, record	A person or place participating in research and has one or more piece of data collected on them.
Syntax	Code, program, script	Programming statements written in a text editor. The statements are machine-readable instructions processed by your computer.

(*Continued*)

Term	Other Terms	Definition
Tool	Platform	A means used to collect data using an instrument (e.g., a paper form, an online survey platform)
Treatment	NA	The individual or group receives the intervention.
Unique participant identifier	Study ID, site ID, unique identifier (UID), subject ID, participant code, record ID	This is a unique numeric or alphanumeric identifier, assigned to every participant or site, and used to create confidential and de-identified data. These identifiers allow researchers to link data across time or measure.
Variable	Column, field, question, data element	Any phenomenon you are collecting information on/trying to measure. These variables will make up columns in your datasets or databases.
Variable name	Header	A shortened symbolic name given the variable in your data to represent the information it contains.
Wave	Time period, time point, event, session	Intervals of data collection over time.

Appendix

For summary purposes, this appendix provides a high-level overview of some of the most common data management activities that occur in each phase of the research life cycle.

DMP (Chapter 5)

- Review oversight requirements.
- Create data sources catalog.
- Create data management plan.

Planning (Chapter 6)

- Develop style guide (see Chapter 9 for more information).
- Choose storage locations (see Chapter 13).
    - Build directory structures for electronic data and physical folder structures for paper data.
- Organize a data management working group (DMWG).
    - Schedule planning meetings and keep meeting notes.
        - Review checklists.
        - Develop workflows.
- Create data collection timeline.
- Assign roles and responsibilities.
- Initiate any necessary processes (e.g., data request, Institutional Review Board application).

Document (Chapter 8)

- Create research protocol.
- Create SOPs.
    - ID schema
    - Consent process
    - Inclusion/exclusion criteria
    - Data collection processes
    - Data entry procedures
- Create data dictionaries.
- Start data cleaning plans.

Create instruments (Chapters 10 and 11)

- Create participant tracking database.
- Create data collection instruments using quality assurance practices.
- Develop consent forms and include data sharing language.

Collect data (Chapter 11)

- Implement quality control procedures during collection.

Track data (Chapter 10 and 11)

- Track incoming data daily in a participant tracking database.

Capture data (Chapter 12)

- Capture original paper and electronic data.
- Capture external data as needed.

Store (Chapter 13)

- Store all raw data and documents for project using secure procedures.

Clean and Validate (Chapter 14)

- Clean data following standardized checklist and data cleaning plans.
- Validate data (e.g., create a codebook to check for errors).
- Store clean data.
- Create participant flow diagram (see Section 8.2.6 for more information).

Version (Chapter 14)

- Version finalized data if errors are found and update changelog.

Prepare to archive (Chapter 15)

- Prepare paper and electronic data for long-term storage.
- Update data inventory with new project datasets (see Section 8.1.4 for more information).
- Develop an internal data reuse process.
- Prepare data and documentation for open sharing (more information in Chapter 16).

Sharing (Chapter 16)

- Share data and documentation in an open repository, using controlled access as needed.

References

Aboumatar, Hanan, Carol Thompson, Emmanuel Garcia-Morales, Ayse P. Gurses, Mohammad Naqibuddin, Jamia Saunders, Samuel W. Kim, and Robert AWise. 2021. "Perspective on Reducing Errors in Research." *Contemporary Clinical Trials Communications* 23 (August): 100838. https://doi.org/10.1016/j.conctc.2021.100838

Aczel, Balazs. 2023. "A Crowdsourced Effort to Develop a Lab Manual Template." *Google Docs*. https://docs.google.com/document/d/1LqGdtHg0dMbj9lsCnC1QOoWzIsnSNRTSek6i3Kls2Ik

Alexander, Rohan. 2023. *Telling Stories with Data*. https://tellingstorieswithdata.com/

Alston, Jesse M., and Jessica A. Rick. 2021. "A Beginner's Guide to Conducting Reproducible Research." *The Bulletin of the Ecological Society of America* 102 (2): e01801. https://doi.org/10.1002/bes2.1801

Arndt, Aaron D., John B. Ford, Barry J. Babin, and Vinh Luong. 2022. "Collecting Samples from Online Services: How to Use Screeners to Improve Data Quality." *International Journal of Research in Marketing* 39 (1): 117–133. https://doi.org/10.1016/j.ijresmar.2021.05.001

Arslan, Ruben C. 2019. "How to Automatically Document Data with the Codebook Package to Facilitate Data Reuse." *Advances in Methods and Practices in Psychological Science* 2 (2): 169–187. https://doi.org/10.1177/2515245919838783

Ashcraft, Alvin. 2022. "Naming Files, Paths, and Namespaces." *Microsoft Build*. https://learn.microsoft.com/en-us/windows/win32/fileio/naming-a-file

Baker, Monya. 2016. "1,500 Scientists Lift the Lid on Reproducibility." *Nature* 533 (7604): 452–454. https://doi.org/10.1038/533452a

Barchard, Kimberly A., Andrew J. Freeman, Elizabeth Ochoa, and Amber K. Stephens. 2020. "Comparing the Accuracy and Speed of Four Data-Checking Methods." *Behavior Research Methods* 52 (1): 97–115. https://doi.org/10.3758/s13428-019-01207-3

Beals, Laura, and Noah Schectman. 2014. "Data Formatting for Performance Management Systems." *AEA 365*. https://aea365.org/blog/laura-beals-and-noah-schectman-on-data-formatting-for-performance-management-systems/

Beaudry, Jennifer, Donna Chen, Bryan Cook, Timothy Errington, Laura Fortunato, Lisa Given, Krystal Hahn, et al. 2022. "The Open Scholarship Survey (OSS)," OSF. https://doi.org/10.17605/OSF.IO/NSBR3

Bennett, L. Michelle, and Howard Gadlin. 2012. "Collaboration and Team Science: From Theory to Practice." *Journal of Investigative Medicine: The Official Publication of the American Federation for Clinical Research* 60 (5): 768–775. https://doi.org/10.231/JIM.0b013e318250871d

Bergmann, Christina. 2023. "The Buffet Approach to Open Science." *CogTales*. https://cogtales.wordpress.com/2023/04/16/the-buffet-approach-to-open-science

Berry, Bailey. 2022. "Guides: Managing Data Sets: Metadata Standards." *Pepperdine Libraries*. https://infoguides.pepperdine.edu/c.php?g=287502&p=1916453

BIDS-Contributors. 2022. "The Brain Imaging Data Structure (BIDS) Specification," Zenodo. https://doi.org/10.5281/ZENODO.3686061

Bishop, Dorothy. 2014. "Data Sharing: Exciting but Scary." *BishopBlog*. http://deevybee.blogspot.com/2014/05/data-sharing-exciting-but-scary.html

Bolam, Mike. 2022. "Guides: Metadata & Discovery @ Pitt: Metadata Standards." *University of Pittsburgh Library System*. https://pitt.libguides.com/metadatadiscovery/metadata-standards

Borer, Elizabeth T., Eric W. Seabloom, Matthew B. Jones, and Mark Schildhauer. 2009. "Some Simple Guidelines for Effective Data Management." *Bulletin of the Ecological Society of America* 90 (2): 205–214. https://doi.org/10.1890/0012-9623-90.2.205

Borghi, John, and Ana Van Gulick. 2021. "Data Management and Sharing: Practices and Perceptions of Psychology Researchers." *PLOS One* 16 (5): e0252047. https://doi.org/10.1371/journal.pone.0252047

Borghi, John, and Ana Van Gulick. 2022. "Promoting Open Science through Research Data Management." *Harvard Data Science Review*, July. https://doi.org/10.1162/99608f92.9497f68e

Borycz, Joshua. 2021. "Implementing Data Management Workflows in Research Groups through Integrated Library Consultancy." *Data Science Journal* 20 (1): 9. https://doi.org/10.5334/dsj-2021-009

Bourgeois, David. 2014. *Information Systems for Business and Beyond*. Published through the Open Textbook Challenge by the Saylor Academy. https://pressbooks.pub/bus206/

Boykis, Vicki. 2021. "The Ghosts in the Data." *Vicki Boykis*. https://veekaybee.github.io/2021/03/26/the-ghosts-in-the-data/

Briney, Kristin. 2015. *Data Management for Researchers: Organize, Maintain and Share Your Data for Research Success*. Research Skills Series. Exeter, UK: Pelagic Publishing.

Briney, Kristin, Heather Coates, and Abigail Goben. 2020. "Foundational Practices of Research Data Management." *Research Ideas and Outcomes* 6 (July): e56508. https://doi.org/10.3897/rio.6.e56508

Broman, Karl W., and Kara H. Woo. 2018. "Data Organization in Spreadsheets." *The American Statistician* 72 (1): 2–10. https://doi.org/10.1080/00031305.2017.1375989

Buchanan, Erin M., Sarah E. Crain, Ari L. Cunningham, Hannah R. Johnson, Hannah Stash, Marietta Papadatou-Pastou, Peder M. Isager, Rickard Carlsson, and Balazs Aczel. 2021. "Getting Started Creating Data Dictionaries: How to Create a Shareable Data Set." *Advances in Methods and Practices in Psychological Science* 4 (1). https://doi.org/10.1177/2515245920928007

Burnard, Lou. 2014. *What Is the Text Encoding Initiative?: How to Add Intelligent Markup to Digital Resources*. OpenEdition Press. https://doi.org/10.4000/books.oep.426

Butters, Oliver W., Rebecca C. Wilson, and Paul R. Burton. 2020. "Recognizing, Reporting and Reducing the Data Curation Debt of Cohort Studies." *International Journal of Epidemiology* 49 (4): 1067–1074. https://doi.org/10.1093/ije/dyaa087

Cakici, Tatiana Baquero. 2017. "Folders v. Metadata in SharePoint Document Libraries." *Enterprise Knowledge*. https://enterprise-knowledge.com/folders-v-metadata-sharepoint-document-libraries/

Campos-Varela, Isabel, and Alberto Ruano-Raviña. 2019. "Misconduct as the Main Cause for Retraction. A Descriptive Study of Retracted Publications and Their Authors." *Gaceta Sanitaria* 33 (4): 356–360. https://doi.org/10.1016/j.gaceta.2018.01.009

Carroll, Stephanie Russo, Ibrahim Garba, Oscar L. Figueroa-Rodríguez, Jarita Holbrook, Raymond Lovett, Simeon Materechera, Mark Parsons, et al. 2020.

"The CARE Principles for Indigenous Data Governance." *Data Science Journal* 19 (1): 43. https://doi.org/10.5334/dsj-2020-043

Castañeda, R. Andrés. 2019. "*A Nice Template for Your Do-Files: Dotemplate.*" *R.Andrés Castañeda Aguilar*. https://randrescastaneda.rbind.io/post/dotemplate/

CDISC. 2023. "CDISC Standards in the Clinical Research Process." *CDISC*. https://www.cdisc.org/standards

Center for Open Science. 2023. "Creating a Data Management Plan (DMP) Document." *OSF Support*. https://help.osf.io/article/144-creating-a-data-management-plan-dmp-document

CESSDA Training Team. 2017. *CESSDA Data Management Expert Guide*. Bergen, Norway: CESSDA ERIC. https://dmeg.cessda.eu/

Ceviren, A. Busra, and Jessica Logan. 2022. "Data Management in Education Research." *Presentation. figshare*. https://doi.org/10.6084/m9.figshare.19514368.v1

Chen, Lu. 2022. "Description of the database normalization basics." *Microsoft Build*. https://learn.microsoft.com/en-us/office/troubleshoot/access/database-normalization-description

Cofield, Melanie. 2023. "LibGuides: Metadata Basics: Key Concepts." *University of Texas Libraries*. https://guides.lib.utexas.edu/metadata-basics/key-concepts

Cohen, Louis, Lawrence Manion, and Keith Morrison. 2007. *Research Methods in Education*. 0th Ed. Routledge. https://doi.org/10.4324/9780203029053

Cook, Bryan G., Jesse Irvan Fleming, Sara Ann Hart, Kathleen Lynne Lane, William Therrien, Wilhelmina Van Dijk, and Sarah Emily Wilson. 2021. "A How-to Guide for Open-Science Practices in Special Education Research." Preprint. EdArXiv. https://doi.org/10.35542/osf.io/zmeba

Cruse, Trisha. 2011. "Best Practice: Provide Budget Information for Your Data Management Plan." *DataOne*. https://dataoneorg.github.io/Education/bestpractices/provide-budget-information

Dahdul, Wasila. 2023. "Research Guides: Research Data Management: Describing Data." *UCI Libraries*. https://guides.lib.uci.edu/datamanagement/describe

Danish National Forum for Research Data Management. 2023. "Metadata." *How to FAIR*. https://howtofair.dk/how-to-fair/metadata/

DeCoster, Jamie. 2023. "Systematic Data Validation." *Prezi.com*. https://prezi.com/view/oOXBw0bPmlReD3T7LZJm/

DIME Analytics. 2021a. "Data Quality Assurance Plan." *The World Bank*. https://dimewiki.worldbank.org/Data_Quality_Assurance_Plan

DIME Analytics. 2021b. "Survey Pilot." *The World Bank*. https://dimewiki.worldbank.org/Survey_Pilot

Doucette, Lise, and Bruce Fyfe. 2013. "Drowning in Research Data: Addressing Data Management Literacy of Graduate Students." https://docplayer.net/8853333-Drowning-in-research-data-addressing-data-management-literacy-of-graduate-students.html

Douglas, Benjamin D., Patrick J. Ewell, and Markus Brauer. 2023. "Data Quality in Online Human-Subjects Research: Comparisons between MTurk, Prolific, CloudResearch, Qualtrics, and SONA." *PLOS One* 18 (3): e0279720. https://doi.org/10.1371/journal.pone.0279720

Duru, Maya, Sarah Kopper, and James Turitto. 2021. "Grant and Budget Management." *The Abdul Latif Jameel Poverty Action Lab (J-PAL)*. https://www.povertyactionlab.org/resource/grant-and-budget-management

Duru, Maya, and Anja Sautmann. 2023. "Institutional Review Board (IRB) Proposals." *The Abdul Latif Jameel Poverty Action Lab (J-PAL)*. https://www.povertyactionlab.org/resource/institutional-review-board-irb-proposals

Eaker, C. 2016. "What Could Possibly Go Wrong? The Impact of Poor Data Management." In Federer, L. (Ed.). *The Medical Library Association's Guide to Data Management for Librarians*. Lanham, Maryland: Rowman; Littlefield Publishing Group. https://trace.tennessee.edu/cgi/viewcontent.cgi?article=1023&context=utk_libpub

Elff, Martin. 2023. "Memisc: Management of Survey Data and Presentation of Analysis Results." https://cran.r-project.org/package=memisc

Elgabry, Omar. 2019. "The Ultimate Guide to Data Cleaning." *Medium*. https://towardsdatascience.com/the-ultimate-guide-to-data-cleaning-3969843991d4

Elsevier Author Services. 2021. "Research Team Structure, Elsevier Language Services Blog." *Elsevier Author Services – Articles*. https://scientific-publishing.webshop.elsevier.com/research-process/research-team-structure/

European Commission. Directorate-General for Research & Innovation. 2016. "H2020 Programme Guidelines on FAIR Data Management in Horizon 2020." *Version3.0*.https://ec.europa.eu/research/participants/data/ref/h2020/grants_manual/hi/oa_pilot/h2020-hi-oa-data-mgt_en.pdf

Farewell, Timothy S. 2018. "My Easy R Script Header Template." *Time Farewell*. https://timfarewell.co.uk/my-r-script-header-template/

Feeney, Laura, Jason Bauman, Julia Chabrier, Geeti Mehra, and Michelle Woodford. 2021. "Using Administrative Data for Randomized Evaluations." *The Abdul Latif Jameel Poverty Action Lab (J-PAL)*. https://www.povertyactionlab.org/resource/using-administrative-data-randomized-evaluations

Feeney, Laura, Sarah Kopper, and Anja Sautmann. 2022. "Ethical Conduct of Randomized Evaluations." *The Abdul Latif Jameel Poverty Action Lab (J-PAL)*. https://www.povertyactionlab.org/resource/ethical-conduct-randomized-evaluations

Figshare. 2023. "Figshare Metadata Schema Overview." *figshare*. https://help.figshare.com/article/figshare-metadata-schema-overview

Filip, Alena. 2023. "San Jose State University Institutional Review Board: Data Management Handbook for Human Subjects Research." https://www.sjsu.edu/research/docs/irb-data-management-handbook.pdf

Foster, Erin D., and Ariel Deardorff. 2017. "Open Science Framework (OSF)." *Journal of the Medical Library Association: JMLA* 105 (2): 203–206. https://doi.org/10.5195/jmla.2017.88

Frederick, Jennifer. 2020. *"Four Strategies for Crafting Inclusive and Effective Demographic Questions." Ithaka S+R*. https://sr.ithaka.org/blog/four-strategies-for-crafting-inclusive-and-effective-demographic-questions/

Frith, Uta. 2020. "Fast Lane to Slow Science." *Trends in Cognitive Sciences* 24 (1): 1–2. https://doi.org/10.1016/j.tics.2019.10.007

Fuchs, Siiri, and Mari Elisa Kuusniemi. 2018. "Making a Research Project Understandable - Guide for Data Documentation," Zenodo. https://doi.org/10.5281/zenodo.1914401

Gaddy, Marcus, and Kassie Scott. 2020. "Principles for Advancing Equitable Data Practice." *Urban Institute*. https://www.urban.org/sites/default/files/publication/102346/principles-for-advancing-equitable-data-practice_0.pdf

Gammons, Rachel. 2022. "Research Guides: Diversity, Equity, and Inclusion in Research: Data Equity." *University of Maryland*. https://lib.guides.umd.edu/ResearchEquity/DataEquity

Ganley, Colleen M., Robert C. Schoen, Mark LaVenia, and Amanda M. Tazaz. 2019. "The Construct Validation of the Math Anxiety Scale for Teachers." *AERA Open* 5 (1): 233285841983970. https://doi.org/10.1177/2332858419839702

Garfinkel, Simson L. 2015. *"De-Identification of Personal Information."* NIST IR 8053. National Institute of Standards; Technology. https://doi.org/10.6028/NIST.IR.8053

Gentzkow, Matthew, and Jesse Shapiro. 2014. "Code and Data for the Social Sciences: A Practitioner's Guide." https://web.stanford.edu/~gentzkow/research/CodeAndData.pdf

Ghent University. 2023. "Sharing Data." *Universiteit Gent*. https://www.ugent.be/en/research/datamanagement/after-research/sharing.htm

Gibson, Michael. 2021. "Data Quality Checks." *The Abdul Latif Jameel Poverty Action Lab (J-PAL)*. https://www.povertyactionlab.org/resource/data-quality-checks

Gibson, Michael, and Wim Louw. 2020. "Survey Programming." *The Abdul Latif Jameel Poverty Action Lab (J-PAL)*. https://www.povertyactionlab.org/resource/survey-programming

Gilmore, Rick O., Joy Lorenzo Kennedy, and Karen E. Adolph. 2018. "Practical Solutions for Sharing Data and Materials from Psychological Research." *Advances in Methods and Practices in Psychological Science* 1 (1): 121–130. https://doi.org/10.1177/2515245917746500

Gonzales, Sara, Matthew B. Carson, and Kristi Holmes. 2022. "Ten Simple Rules for Maximizing the Recommendations of the NIH Data Management and Sharing Plan." *PLOS Computational Biology* 18 (8): e1010397. https://doi.org/10.1371/journal.pcbi.1010397

Goodman, Alyssa, Alberto Pepe, Alexander W. Blocker, Christine L. Borgman, Kyle Cranmer, Merce Crosas, Rosanne Di Stefano, et al. 2014. "Ten Simple Rules for the Care and Feeding of Scientific Data." *PLOS Computational Biology* 10 (4): e1003542. https://doi.org/10.1371/journal.pcbi.1003542

Gower-Page, Craig, and Kieran Martin. 2020. "Diffdf: Dataframe Difference Tool." https://cran.r-project.org/web/packages/diffdf/index.html

Grace-Martin, Karen. 2013. "The Wide and Long Data Format for Repeated Measures Data." *The Analysis Factor*. https://www.theanalysisfactor.com/wide-and-long-data/

Grave, Joana. 2021. "Scientists Should Be Open about Their Mistakes." *Nature Human Behaviour* 5 (12): 1593. https://doi.org/10.1038/s41562-021-01225-2

Grynoch, Tess. 2024. "Resource Guides: Research Data Management Resources: Data Documentation." *Lamar Soutter Library*. https://libraryguides.umassmed.edu/research_data_management_resources/documentation

Gueguen, Gretchen. 2023. "New OSF Metadata to Support Data Sharing Policy Compliance." *Center for Open Science*. https://www.cos.io/blog/new-osf-metadata-to-support-data-sharing-policy-compliance

Hansen, Karsten Kryger. 2017. "DataFlowToolkit.dk." https://doi.org/10.5278/16k4-4n24

Hardwicke, Tom Elis, Maya B. Mathur, Kyle Earl MacDonald, Gustav Nilsonne, George Christopher Banks, Mallory Kidwell, Alicia Hofelich Mohr, et al. 2018. "Data Availability, Reusability, and Analytic Reproducibility: Evaluating the Impact of a Mandatory Open Data Policy at the Journal Cognition." Preprint. MetaArXiv. https://doi.org/10.31222/osf.io/39cfb

Hart, Sara, Chris Schatschneider, and Jeanette Taylor. 2021. "Florida Twin Project on Reading, Behavior, and Environment." LDbase. https://doi.org/10.33009/ldbase.1624451991.3667

Harvard Longwood Research Data Management. 2023. "File Naming Conventions." *Longwood Research Data Management*. https://datamanagement.hms.harvard.edu/plan-design/file-naming-conventions

Hayslett, Michele. 2022. "LibGuides: Metadata for Data Management: A Tutorial: Basic Elements." *UNC University Libraries*. https://guides.lib.unc.edu/metadata/basic-elements

Helm, Shirley. 2022. "Cultivating an Effective Research Team through Application of Team Science Principles." *SOCRA Blog*. https://www.socra.org/blog/cultivating-an-effective-research-team-through-application-of-team-science-principles/

Henry, Teague. 2021. "Data Management for Researchers: Three Tales." *Teague Henry*. https://www.teaguehenry.com/strings-not-factors/2021/1/24/data-management-for-researchers-three-terrifying-tales

Holdren, John. 2013. "OSTP Memo: 'Increasing Access to the Results of Federally Funded Scientific Research'." https://obamawhitehouse.archives.gov/sites/default/files/microsites/ostp/ostp_public_access_memo_2013.pdf

Hollmann, Susanne, Marcus Frohme, Christoph Endrullat, Andreas Kremer, Domenica D'Elia, Babette Regierer, and Alina Nechyporenko. 2020. "Ten Simple Rules on How to Write a Standard Operating Procedure." *PLoS Computational Biology* 16 (9): e1008095. https://doi.org/10.1371/journal.pcbi.1008095

Houtkoop, Bobby Lee, Chris Chambers, Malcolm Macleod, Dorothy V. M. Bishop, Thomas E. Nichols, and Eric-Jan Wagenmakers. 2018. "Data Sharing in Psychology: A Survey on Barriers and Preconditions." *Advances in Methods and Practices in Psychological Science* 1 (1): 70–85. https://doi.org/10.1177/2515245917751886

Hoyt, Peter, Maneesha Sane, Aleksandra Nenadic, Christie Bahlai, Katrin Leinweber, Tracy Teal, Ileana Fenwick, et al. 2023. "Datacarpentry/Spreadsheet-Ecology-Lesson: Data Carpentry: Data Organization in Spreadsheets for Ecologists (2023.05)." *Zenodo*. https://doi.org/10.5281/ZENODO.7892279

Hubbard, Aleata. 2017. "Data Cleaning in Mathematics Education Research: The Overlooked Methodological Step." https://eric.ed.gov/?id=ED583982

ICPSR. 2011. *Guide to Codebooks 1st Edition*. Ann Arbor, MI. https://www.icpsr.umich.edu/files/deposit/Guide-to-Codebooks_v1.pdf

ICPSR. 2020. *Guide to Social Science Data Preparation and Archiving: 6th Ed*. https://www.icpsr.umich.edu/files/deposit/dataprep.pdf

ICPSR. 2022. "An Introduction to Common Data Elements." https://www.youtube.com/watch?v=GsnoiPzxC4g

ICPSR. 2023a. "ICPSR, Data Management, Metadata." *ICPSR* https://www.icpsr.umich.edu/web/pages/datamanagement/lifecycle/metadata.html

ICPSR. 2023b. "Start Sharing Data Deposit." *ICPSR* https://www.icpsr.umich.edu/web/pages/deposit/index.html

ICPSR. 2023c. "When Should I Deposit My Data with ICPSR?" *ICPSR* https://www.icpsr.umich.edu/web/ICPSR/cms/3944

Institute of Education Sciences. 2019. "SLDS Topical Webinar Summary: Managing Data Requests." https://slds.ed.gov/services/PDCService.svc/GetPDCDocumentFile?fileId=34570

Institute of Education Sciences. 2022. "Standards for Excellence in Education Research." *Institute of Education Sciences*. https://ies.ed.gov/seer/index.asp

Institute of Education Sciences. 2023a. "Frequently Asked Questions about Providing Public Access to Data." *Resources for Researchers*. https://ies.ed.gov/funding/datasharing_faq.asp

Institute of Education Sciences. 2023b. "Policy Statement on Public Access to Data Resulting from IES Funded Grants." *Resources for Researchers*. https://ies.ed.gov/funding/datasharing_policy.asp

International Organization for Standardization. 2017. "ISO 8601 — Date and Time Format." *ISO*. https://www.iso.org/iso-8601-date-and-time-format.html

IPUMS USA. 2023. "Introduction to Data Editing and Allocation." *IPUMS USA*. https://usa.ipums.org/usa/flags.shtml

Jørgensen, Carsten Krogh, and Bo Karlsmose. 1998. "Validation of Automated Forms Processing." *Computers in Biology and Medicine* 28 (6): 659–667. https://doi.org/10.1016/S0010-4825(98)00038-9

Kaplowitz, Rella, and Jasmine Johnson. 2020. "5 Best Practices for Equitable and Inclusive Data Collection." *Schusterman Family Philanthropies*. https://www.schusterman.org/article/5-best-practices-for-equitable-and-inclusive-data-collection

Kathawalla, Ummul-Kiram, Priya Silverstein, and Moin Syed. 2021. "Easing into Open Science: A Guide for Graduate Students and Their Advisors." *Collabra: Psychology* 7 (1): 18684. https://doi.org/10.1525/collabra.18684

Klein, Olivier, Tom E. Hardwicke, Frederik Aust, Johannes Breuer, Henrik Danielsson, Alicia Hofelich Mohr, Hans IJzerman, Gustav Nilsonne, Wolf Vanpaemel, and Michael C. Frank. 2018. "A Practical Guide for Transparency in Psychological Science." Edited by Michéle Nuijten and Simine Vazire. *Collabra: Psychology* 4 (1): 20. https://doi.org/10.1525/collabra.158

Kline, Melissa, Tal Yakori, Lisa DeBruine, Alicia H. Mohr, Erin M. Buchanan, R. Arslan, Stephen R. Martin, et al. 2018. "A Technical Specification for Psychological Datasets." *Google Docs*. https://docs.google.com/document/d/1u8o5jnWk0Iqp_J06PTu5NjBfVsdoPbBhstht6W0fFp0

Koos, Jessica. 2023. "Research & Subject Guides: Research Data Guide: Data Collection and Creation." https://guides.library.stonybrook.edu/research-data/collection

Kopper, Sarah, and Katie Parry. 2023a. "Questionnaire Piloting." *The Abdul Latif Jameel Poverty Action Lab (J-PAL)*. https://www.povertyactionlab.org/resource/questionnaire-piloting

Kopper, Sarah, and Katie Parry. 2023b. "Survey Design." *The Abdul Latif Jameel Poverty Action Lab (J-PAL)*. https://www.povertyactionlab.org/resource/survey-design

Kopper, Sarah, Anja Sautmann, and James Turitto. 2023a. "Data de-Identification." *The Abdul Latif Jameel Poverty Action Lab (J-PAL)*. https://www.povertyactionlab.org/resource/data-de-identification

Kopper, Sarah, Anja Sautmann, and James Turitto. 2023b. "Data Publication." *The Abdul Latif Jameel Poverty Action Lab (J-PAL).* https://www.povertyactionlab.org/resource/data-publication

Kovacs, Marton, Rink Hoekstra, and Balazs Aczel. 2021. "The Role of Human Fallibility in Psychological Research: A Survey of Mistakes in Data Management." *Advances in Methods and Practices in Psychological Science* 4 (4): 251524592110459. https://doi.org/10.1177/25152459211045930

Krishna, Vamsi. 2018. "How to Tag Files in Windows for Easy Retrieval." *Make Tech Easier.* https://www.maketecheasier.com/tag-files-in-windows/

Kush, R. D., D. Warzel, M. A. Kush, A. Sherman, E. A. Navarro, R. Fitzmartin, F. Pétavy, et al. 2020. "FAIR Data Sharing: The Roles of Common Data Elements and Harmonization." *Journal of Biomedical Informatics* 107 (July): 103421. https://doi.org/10.1016/j.jbi.2020.103421

Laskowski, Kate. 2020. "What to Do When You Don't Trust Your Data Anymore – Laskowski Lab at UC Davis." *Laskowski Lab at UC Davis.* https://laskowskilab.faculty.ucdavis.edu/2020/01/29/retractions/

LDbase. 2023. "Information to Gather before Uploading Your Data." *LDbase.* https://www.ldbase.org/resources/user-guide/information-to-gather

LeClere, Felicia. 2010. "Too Many Researchers Are Reluctant to Share Their Data." *The Chronicle of Higher Education*, August. http://www.chronicle.com/article/Too-Many-Researchers-Are/123749

Levenstein, Margaret C., and Jared A. Lyle. 2018. "Data: Sharing Is Caring." *Advances in Methods and Practices in Psychological Science* 1 (1): 95–103. https://doi.org/10.1177/2515245918758319

Levesque, Karen, Robert Fitzgerald, and Jay Pfeiffer. 2015. "A Guide to Using State Longitudinal Data for Applied Research (NCEE 2015–4013)." Washington, DC: U.S. Department of Education, Institute of Education Sciences, National Center for Education Evaluation; Regional Assistance, Analytic Technical Assistance; Development. https://ies.ed.gov/ncee/rel/regions/central/pdf/CE5.3.2-A-Guide-to-Using-State-Longitudinal-Data-for-Applied-Research.pdf

Lewis, Crystal. 2022a. "Using a Data Dictionary as Your Roadmap to Quality Data." *Crystal Lewis.* https://cghlewis.com/blog/data_dictionary/

Lewis, Crystal. 2022b. "How to Export Analysis-Ready Survey Data." *Crystal Lewis.* https://cghlewis.com/blog/survey_data/

Logan, Jessica. 2021. "Data Sharing and Data Shared." *Presentation. figshare.* https://doi.org/10.6084/m9.figshare.15040740.v1

Logan, Jessica, and Sara Hart. 2023. "Within & Between S4e2." *Within & Between.* http://www.withinandbetweenpod.com/

Logan, Jessica, Sara Hart, and Christopher Schatschneider. 2021. "Data Sharing in Education Science." *AERA Open*, 7. https://doi.org/10.1177/23328584211006475

Lucidchart. 2019. "What Is a Workflow? Benefits and Examples of Repeatable Processes." *Lucidchart Blog.* https://www.lucidchart.com/blog/what-is-workflow

Macquarie University. 2023. "Research Data Sensitivity, Security and Storage Guideline." *Macquarie University.* https://policies.mq.edu.au/document/view.php?id=302

Malow, Beth A., Anjalee Galion, Frances Lu, Nan Kennedy, Colleen E. Lawrence, Alison Tassone, Lindsay O'Neal, et al. 2021. "A REDCap-Based Model for Online Interventional Research: Parent Sleep Education in Autism." *Journal of Clinical and Translational Science* 5 (1): e138. https://doi.org/10.1017/cts.2021.798

Markowetz, Florian. 2015. "Five Selfish Reasons to Work Reproducibly." *Genome Biology* 16 (1): 274. https://doi.org/10.1186/s13059-015-0850-7

Mathematica. 2023. "Tips for Conducting Equitable and Culturally Responsive Research." *Mathematica*. https://www.mathematica.org/features/tips-for-conducting-equitable-and-culturally-responsive-evaluation

McKay Bowen, Claire, and Joshua Snoke. 2023. "Do No Harm Guide: Applying Equity Awareness in Data Privacy Methods." *Urban Institute*. https://www.urban.org/sites/default/files/2023-03/Applying%20Equity%20Awareness%20In%20Data%20Privacy%20Methods_0.pdf

McKenzie, Patrick. 2010. "Falsehoods Programmers Believe about Names." *Kalzumeus Software*. https://www.kalzumeus.com/2010/06/17/falsehoods-programmers-believe-about-names/

Mehr, Samuel. 2020. "How to… Write a Lab Handbook." *RSB*. https://www.rsb.org.uk//biologist-features/how-to-write-a-lab-handbook

Meyer, Michelle N. 2018. "Practical Tips for Ethical Data Sharing." *Advances in Methods and Practices in Psychological Science* 1 (1): 131–144. https://doi.org/10.1177/2515245917747656

Michener, William K. 2015. "Ten Simple Rules for Creating a Good Data Management Plan." *PLOS Computational Biology* 11 (10): e1004525. https://doi.org/10.1371/journal.pcbi.1004525

Michigan State University. 2023. "Budgeting for Data Management." *Michigan State University*. https://spa.msu.edu/PL/Portal/3700/BudgetingforDataManagement

Microsoft. 2023. "Restrictions and Limitations in OneDrive and SharePoint - Microsoft Support." *Microsoft*. https://support.microsoft.com/en-us/office/restrictions-and-limitations-in-onedrive-and-sharepoint-64883a5d-228e-48f5-b3d2-eb39e07630fa

Miranda, Dana, and Rob Watts. 2022. "What Is A RACI Chart? How This Project Management Tool Can Boost Your Productivity." *Forbes Advisor*. https://www.forbes.com/advisor/business/raci-chart/

Mohr, Alicia Hofelich, Jake Carlson, Lizhao Ge, Joel Herndon, Wendy Kozlowski, Jennifer Moore, Jonathan Petters, Shawna Taylor, and Cynthia Hudson Vitale. 2024. "Making Research Data Publicly Accessible: Estimates of Institutional & Researcher Expenses." *Association of Research Libraries*. https://doi.org/10.29242/report.radsexpense2024

Mons, Barend. 2020. "Invest 5% of Research Funds in Ensuring Data Are Reusable." *Nature* 578 (7796): 491–491. https://doi.org/10.1038/d41586-020-00505-7

Morehouse, Kirsten, Benedek Kurdi, and Brian A. Nosek. 2023. "Responsible Data Sharing: Identifying and Remedying Possible Re-Identification of Human Participants." Preprint. MetaArXiv. https://doi.org/10.31222/osf.io/5m3cx

National Archive of Criminal Justice Data. 2023. "Guidelines for Depositing NIJ and OJJDP Data at NACJD." https://www.icpsr.umich.edu/files/NACJD/pdf/Guidelines_for_Depositing_NIJ_and_OJJDP_Data_at_NACJD.pdf

National Endowment for the Humanities. 2018. "Data Managment Plans for NEH Office of Digital Humanities Proposals and Awards." https://www.neh.gov/sites/default/files/2018-06/data_management_plans_2018.pdf

National Institute of Justice. 2007. "The 'Common Rule'." *National Institute of Justice*. https://nij.ojp.gov/funding/common-rule

National Institute of Mental Health. 2023. "Notice of Data Sharing Policy for the National Institute of Mental Health." NOT-MH-23-100. https://grants.nih.gov/grants/guide/notice-files/NOT-MH-23-100.html

National Institutes of Health. 2021. "Update on Implementation of New NIH Data Management and Sharing Policy." https://www.youtube.com/watch?v=ZcD641ptWIY

National Institutes of Health. 2022. "Supplemental Information to the NIH Policy for Data Management and Sharing: Responsible Management and Sharing of American Indian/Alaska Native Participant Data." NOT-OD-22-214. https://grants.nih.gov/grants/guide/notice-files/NOT-OD-22-214.html

National Institutes of Health. 2023a. "Budgeting for Data Management & Sharing Data Sharing." *NIH Scientific Data Sharing*. https://sharing.nih.gov/data-management-and-sharing-policy/planning-and-budgeting-for-data-management-and-sharing/budgeting-for-data-management-sharing#after

National Institutes of Health. 2023b. "Selecting a Data Repository Data Sharing." *NIH Scientific Data Sharing*. https://sharing.nih.gov/data-management-and-sharing-policy/sharing-scientific-data/selecting-a-data-repository#desirable-characteristics-for-all-data-repositories

National Institutes of Health. 2023c. "Final NIH Policy for Data Management and Sharing." NOT-OD-21-013. https://grants.nih.gov/grants/guide/notice-files/NOT-OD-21-013.html

National Science and Technology Council. 2022. "Desirable Characteristics of Data Repositories for Federally Funded Research." *Executive Office of the President of the United States*. https://doi.org/10.5479/10088/113528

National Science Foundation. 2023. "NSF Public Access Plan 2.0." https://nsf-gov-resources.nsf.gov/2023-06/NSF23104.pdf

Neild, R. C., D. Robinson, and J. Agufa. 2022. "Sharing Study Data: A Guide for Education Researchers. (NCEE 2022-004)." *U.S. Department of Education, Institute of Education Sciences, National Center for Education Evaluation and Regional Assistance*. https://ies.ed.gov/ncee/pubs/2022004/pdf/2022004.pdf

Nelson, Alondra. 2022. "OSTP Memo: Ensuring Free, Immediate, and Equitable Access to Federally Funded Research." https://www.whitehouse.gov/wp-content/uploads/2022/08/08-2022-OSTP-Public-Access-Memo.pdf

Northern Illinois University. 2023. "Data Collection." *Responsible Conduct in Data Management* https://ori.hhs.gov/education/products/n_illinois_u/datamanagement/dctopic.html

NUCATS. 2023. "Standard Operating Procedures (SOPs)." https://www.nucats.northwestern.edu/docs/cecd/overview-of-sops.pdf

O'Toole, Elisabeth, Laura Feeney, Kenya Heard, and Rohit Naimpally. 2018. "Data Security Procedures for Researchers." J-PAL North America. https://www.povertyactionlab.org/sites/default/files/Data_Security_Procedures_December.pdf

Office for Human Research Protections. 2009. "Federal Policy for the Protection of Human Subjects ('Common Rule')." Text. *HHS.gov*. https://www.hhs.gov/ohrp/regulations-and-policy/regulations/common-rule/index.html

Office for Human Research Protections. 2016. "*45 CFR 46*." Text. *HHS.gov*. https://www.hhs.gov/ohrp/regulations-and-policy/regulations/45-cfr-46/index.html

Office for Human Research Protections. 2018. "*Revised Common Rule Q&As*." *HHS.gov*. https://www.hhs.gov/ohrp/education-and-outreach/revised-common-rule/revised-common-rule-q-and-a/index.html

OpenAIRE_eu. 2018. "Basics of Research Data Management." https://www.youtube.com/watch?v=3sDhQRIYUmA

Oregon State University. 2012. "What Is the Institutional Review Board (IRB)?" *Oregon State University*. https://research.oregonstate.edu/irb/what-institutional-review-board-irb

Pacific University Oregon. 2014. "Data Security and Storage." *Pacific University*. https://www.pacificu.edu/academics/research/scholarship-and-sponsored-projects/research-compliance-integrity/institutional-review-board/irb-policies-recommended-practices/data-security-storage

Palmer, Brendan. 2023. "Advice from the Data Stewards." *Sonrai*. https://datastewards.ie/advice-from-the-experts-dr-brendan-palmer/

Patridge, Emily F., and Tania P. Bardyn. 2018. "Research Electronic Data Capture (REDCap)." *Journal of the Medical Library Association: JMLA* 106 (1): 142–144. https://doi.org/10.5195/jmla.2018.319

Pew Research Center. 2023. "Writing Survey Questions." *Pew Research Center*. https://www.pewresearch.org/our-methods/u-s-surveys/writing-survey-questions/

Princeton University. 2023a. "Best Practices for Data Analysis of Confidential Data." *Princeton Research*. https://ria.princeton.edu/human-research-protection/data/best-practices-for-data-a

Princeton University. 2023b. *Research Lifecycle Guide: Princeton Research Data Service*. *Princeton Research Data Service* https://researchdata.princeton.edu/research-lifecycle-guide/research-lifecycle-guide

R Core Team. 2023. "R: The R Base Package." *R Foundation for Statistical Computing*. https://stat.ethz.ch/R-manual/R-devel/library/base/html/00Index.html

Renbarger, Rachel, Jill L. Adelson, Joshua Rosenberg, Sondra M Stegenga, Olivia Lowrey, Pamela Rose Buckley, and Qiyang Zhang. 2022. "Champions of Transparency in Education: What Journal Reviewers Can Do to Encourage Open Science Practices." EdArXiv. https://doi.org/10.35542/osf.io/xqfwb

Reynolds, John H., Emily D. Silverman, Kelly Chadbourne, Sean Finn, Richard Kearney, Kaylene Keller, Chris Lett, Socheata Lor, Kristin Shears, and John Swords. 2014. "Data Are 'Trust Resources': A Strategy for Managing the U.S. Fish and Wildlife Serivce's Scientific Data." Report from the {U}.{S}. {Fish} and {Wildlife} {Service}'s ad hoc subcommittee on data management. https://www.arlis.org/docs/vol1/FWS/2014/929063674.pdf

Reynolds, Tara, Christopher Schatschneider, and Jessica Logan. 2022. "The Basics of Data Management." *figshare*. https://doi.org/10.6084/m9.figshare.13215350.v2

Riederer, Emily. 2020. "Column Names as Contracts." *Emily Riederer*. https://emilyriederer.netlify.app/post/column-name-contracts/

Riederer, Emily. 2021. "Make Grouping a First-Class Citizen in Data Quality Checks." *Emily Riederer*. https://emilyriederer.netlify.app/post/grouping-data-quality/

Salfen, Jeremy. 2018. "Building a Data Practice from Scratch." *Locally Optimistic*. https://locallyoptimistic.com/post/building-a-data-practice/

San Martin, Luis Eduardo, Rony Rodriguez-Ramirez, and Mizuhiro Suzuki. 2023. "Stata Linter Produces Stata Code That Sparks Joy." *World Bank Blogs*. https://blogs.worldbank.org/impactevaluations/stata-linter-produces-stata-code-sparks-joy

Schatschneider, Christopher, Ashley Edwards, and Jeffrey Shero. 2021. "De-Identification Guide." *figshare*. https://doi.org/10.6084/m9.figshare.13228664.v2

Schema.org. 2023. "Schema.org." https://www.schema.org/

Schmidt, Carsten Oliver, Stephan Struckmann, Cornelia Enzenbach, Achim Reineke, Jürgen Stausberg, Stefan Damerow, Marianne Huebner, Börge Schmidt, Willi Sauerbrei, and Adrian Richter. 2021. "Facilitating Harmonized Data Quality Assessments. A Data Quality Framework for Observational Health Research Data Collections with Software Implementations in R." *BMC Medical Research Methodology* 21 (1): 63. https://doi.org/10.1186/s12874-021-01252-7

Schmitt, Charles P., and Margaret Burchinal. 2011. "Data Management Practices for Collaborative Research." *Frontiers in Psychiatry* 2 (July): 47. https://doi.org/10.3389/fpsyt.2011.00047

Schoen, Robert C., and Gizem Solmaz-Ratzlaff. 2023. "Guest Post—Concrete Models for Educational Data Sharing." *Center for Open Science*. https://www.cos.io/blog/concrete-models-for-educational-data-sharing

Schulz, Kenneth F., Douglas G. Altman, David Moher, and CONSORT Group. 2010. "CONSORT 2010 Statement: Updated Guidelines for Reporting Parallel Group Randomised Trials." *BMJ (Clinical Research Ed.)* 340 (March): c332. https://doi.org/10.1136/bmj.c332

Seastrom, Marilyn M. 2002. "NCES Statistical Standards." *National Center for Education Statistics*. https://nces.ed.gov/pubsearch/pubsinfo.asp?pubid=2003601

Simone, Melissa. 2019. "How to Battle the Bots Wrecking Your Online Study." *Behavioral Scientist*. https://behavioralscientist.org/how-to-battle-the-bots-wrecking-your-online-study/

Society of Critical Care Medicine. 2018. "Building an Efficient Database for Your Research." https://www.youtube.com/watch?v=9ELr2P2pQZg

Spinks, Madeleine. 2024. "Introducing the Data Cleaning Day." *Data Orchard*. https://www.dataorchard.org.uk/news/introducing-the-data-cleaning-day

Stangroom, Jeremy. 2019. "Rules for Naming Variables in SPSS - Quick SPSS Tutorial." *EZ SPSS Tutorials*. https://ezspss.com/rules-for-naming-variables-in-spss/

Stodden, Victoria, Jennifer Seiler, and Zhaokun Ma. 2018. "An Empirical Analysis of Journal Policy Effectiveness for Computational Reproducibility." *Proceedings of the National Academy of Sciences* 115 (11): 2584–2589. https://doi.org/10.1073/pnas.1708290115

Strand, Julia. 2020. "Scientists Make Mistakes. I Made a Big One." *Elemental*. https://elemental.medium.com/when-science-needs-self-correcting-a130eacb4235

Strand, Julia. 2021. "Error Tight: Exercises for Lab Groups to Prevent Research Mistakes." Preprint. PsyArXiv. https://psyarxiv.com/rsn5y/

Sweeney, Latanya. 2002. "K-Anonymity: A Model for Protecting Privacy." *International Journal of Uncertainty, Fuzziness and Knowledge-Based Systems* 10 (05): 557–570. https://doi.org/10.1142/S0218488502001648

Teitcher, Jennifer E. F., Walter O. Bockting, José A. Bauermeister, Chris J. Hoefer, Michael H. Miner, and Robert L. Klitzman. 2015. "Detecting, Preventing, and Responding to 'Fraudsters' in Internet Research: Ethics and Tradeoffs." *The Journal of Law, Medicine & Ethics: A Journal of the American Society of Law, Medicine & Ethics* 43 (1): 116–133. https://doi.org/10.1111/jlme.12200

Tenopir, Carol, Suzie Allard, Priyanki Sinha, Danielle Pollock, Jess Newman, Elizabeth Dalton, Mike Frame, and Lynn Baird. 2016. "Data Management Education from the Perspective of Science Educators." *International Journal of Digital Curation* 11 (1): 232–251. https://doi.org/10.2218/ijdc.v11i1.389

The National Commission for the Protection of Human Subjects of Biomedical and Behavioral Research. 1979. "The Belmont Report." https://www.hhs.gov/ohrp/sites/default/files/the-belmont-report-508c_FINAL.pdf

The Turing Way Community. 2022. *The Turing Way: A Handbook for Reproducible, Ethical and Collaborative Research*. Zenodo. https://doi.org/10.5281/ZENODO.3233853

The White House. 2013. "Executive Order – Making Open and Machine Readable the New Default for Government Information." *Whitehouse.gov*. https://obamawhitehouse.archives.gov/the-press-office/2013/05/09/executive-order-making-open-and-machine-readable-new-default-government

Thielen, Joanna, and Amanda Nichols Hess. 2017. "Advancing Research Data Management in the Social Sciences: Implementing Instruction for Education Graduate Students into a Doctoral Curriculum." *Behavioral & Social Sciences Librarian* 36 (1): 16–30. https://doi.org/10.1080/01639269.2017.1387739

Towse, Andrea S., David A. Ellis, and John N. Towse. 2021. "Making Data Meaningful: Guidelines for Good Quality Open Data." *The Journal of Social Psychology* 161 (4): 395–402. https://doi.org/10.1080/00224545.2021.1938811

U.S. Department of Health and Human Services. 2012. "Guidance Regarding Methods for De-Identification of Protected Health Information in Accordance with the Health Insurance Portability and Accountability Act (HIPAA) Privacy Rule." Text. https://www.hhs.gov/hipaa/for-professionals/privacy/special-topics/de-identification/index.html

U.S. Department of Health and Human Services. 2018. "What's New in IRB Review under the Revised Common Rule." https://www.youtube.com/watch?v=zDsUUs9j3sQ

UC Merced Library. 2023. "What Is a Data Dictionary? *UC Merced Library*." https://library.ucmerced.edu/data-dictionaries

UK Data Service. 2023. "Research Data Management." *UK Data Service*. https://ukdataservice.ac.uk/learning-hub/research-data-management/

UNC Office of Human Research Ethics. 2020. "De-Identified, Coded, or Anonymous? How Do I Know?" *UNC Research*. https://research.unc.edu/2020/05/01/de-identified-coded-or-anonymous-how-do-i-know/

United States Department of Health and Human Services. Administration for Children and Families, Office of Planning, Research and Evaluation. 2023. "Study of Coaching Practices in Early Care and Education Settings (SCOPE), United States, 2019-2021." Inter-University Consortium for Political; Social Research [distributor]. https://doi.org/10.3886/ICPSR38290.V2

University of Iowa Libraries. 2024. "Metadata." *Iowa University Libraries*. https://www.lib.uiowa.edu/data/share/metadata/

University of Michigan. 2023. "Examples of Sensitive Data by Classification Level / Safecomputing.umich.edu." *U-M Information and Technology Services*.

https://safecomputing.umich.edu/protect-the-u/safely-use-sensitive-data/examples-by-level

University of Washington. 2023. "Sharing Information and Data." *UW Research*. https://www.washington.edu/research/myresearch-lifecycle/setup/collaborations/sharing-information-and-data/

USGS. 2021. "Tools for Creating Metadata Records." *USGS*. https://www.usgs.gov/data-management/metadata-creation#tools

USGS. 2023. "What Are the Differences between Data, a Dataset, and a Database? U.S. Geological Survey." *USGS*. https://www.usgs.gov/faqs/what-are-differences-between-data-dataset-and-database

Valentine, Theresa. 2011. "Best Practice: Define Roles and Assign Responsibilities for Data Management." *DataOne*. https://dataoneorg.github.io/Education/bestpractices/define-roles-and

Van Bochove, Kees, Pinar Alper, and Wei Gu. 2023. "Data Quality." *RDMkit*. https://rdmkit.elixir-europe.org/data_quality

Van Den Eynden, Veerle, Louise Corti, Matthew Woollard, Libby Bishop, and Laurence Horton. 2011. "Managing and Sharing Data." https://dam.ukdataservice.ac.uk/media/622417/managingsharing.pdf

Van Dijk, Wilhelmina, Christopher Schatschneider, and Sara A. Hart. 2021. "Open Science in Education Sciences." *Journal of Learning Disabilities* 54 (2): 139–152. https://doi.org/10.1177/0022219420945267

Veselovsky, Veniamin, Manoel Horta Ribeiro, and Robert West. 2023. "Artificial Artificial Artificial Intelligence: Crowd Workers Widely Use Large Language Models for Text Production Tasks." arXiv. https://doi.org/10.48550/ARXIV.2306.07899

Vines, Timothy H., Arianne Y. K. Albert, Rose L. Andrew, Florence Débarre, Dan G. Bock, Michelle T. Franklin, Kimberly J. Gilbert, Jean-Sébastien Moore, Sébastien Renaut, and Diana J. Rennison. 2014. "The Availability of Research Data Declines Rapidly with Article Age." *Current Biology* 24 (1): 94–97. https://doi.org/10.1016/j.cub.2013.11.014

Washington University in St. Louis. 2023. "Roles and Responsibilities - Research - Washington University in St. Louis." *Research*. https://research.wustl.edu/about/roles-responsibilities/

Webb, Margaret A., and June P. Tangney. 2022. "Too Good to Be True: Bots and Bad Data From Mechanical Turk." *Perspectives on Psychological Science*, November, 174569162211200. https://doi.org/10.1177/17456916221120027

White, Ethan, Elita Baldridge, Zachary Brym, Kenneth Locey, Daniel McGlinn, and Sarah Supp. 2013. "Nine Simple Ways to Make It Easier to (Re)use Your Data." *Ideas in Ecology and Evolution* 6 (2). https://doi.org/10.4033/iee.2013.6b.6.f

Wickham, Hadley. 2014. "Tidy Data." *Journal of Statistical Software* 59 (September): 1–23. https://doi.org/10.18637/jss.v059.i10

Wickham, Hadley. 2021. *Welcome: The Tidyverse Style Guide*. https://style.tidyverse.org/index.html

Wickham, Hadley, Mine Çetinkaya-Rundel, and Garrett Grolemund. 2023. *R for Data Science: Import, Tidy, Transform, Visualize, and Model Data*. 2nd Ed. Beijing, Boston, Farnham, Sebastopol, Tokyo: O'Reilly.

Wilkinson, Mark D., Michel Dumontier, IJsbrand Jan Aalbersberg, Gabrielle Appleton, Myles Axton, Arie Baak, Niklas Blomberg, et al. 2016. "The FAIR Guiding Principles for Scientific Data Management and Stewardship." *Scientific Data* 3 (1): 160018. https://doi.org/10.1038/sdata.2016.18

Wilson, Greg, Jennifer Bryan, Karen Cranston, Justin Kitzes, Lex Nederbragt, and Tracy K. Teal. 2017. "Good Enough Practices in Scientific Computing." *PLOS Computational Biology* 13 (6): e1005510. https://doi.org/10.1371/journal.pcbi.1005510

Yenni, Glenda M., Erica M. Christensen, Ellen K. Bledsoe, Sarah R. Supp, Renata M. Diaz, Ethan P. White, and S. K. Morgan Ernest. 2019. "Developing a Modern Data Workflow for Regularly Updated Data." *PLoS Biology* 17 (1): e3000125. https://doi.org/10.1371/journal.pbio.3000125

Zhou, Xuan, Zhihong Xu, and Ashlynn Kogut. 2023. "Research Data Management Needs Assessment for Social Sciences Graduate Students: A Mixed Methods Study." *PLOS One* 18 (2): e0282152. https://doi.org/10.1371/journal.pone.0282152

Index

Pages in *italics* refer to figures, pages in **bold** refer to tables, and pages followed by n refer to notes.

E

M

N

O

P

Printed in the United States
by Baker & Taylor Publisher Services

Printed in the United States
by Baker & Taylor Publisher Services